DATE DUE

AG		
DE 2'98		
JE 11 01		
MY 20'02		

DEMCO 38-296

STARTING IN HEATING & AIR CONDITIONING SERVICE

BNP Business News Publishing Company
Troy • Michigan

Editor: Joanna Turpin
Production Coordinator: Mark Leibold
Cover Illustration: Bryan Ureel

Technical review performed by Clayton Carrico
Book compiled with the collaboration of Lionel E. LaRocque

Library of Congress Cataloging in Publication Data

Starting in Heating and Air Conditioning Service/
  [technical review performed by Clayton Carrico: book
  compiled with the collaboration of Lionel E. LaRocque].
      p. cm.
    Includes index.
    ISBN 0-912524-63-4
    1. Heating. 2 Air conditioning I. Carrico, Clayton H.,
  1926- . II. LaRocque, Lionel Edward, 1937- .
TH7012.S73   1992   92-11928
697—dc20   CIP
                                    Rev.

Printed in United States
7 6 5 4 3

DISCLAIMER

This book is only considered to be a general guide. The author and publisher have neither liability nor can they be responsible to any person or entity for any misunderstanding, misuse or misapplication that would cause loss or damage of any kind, including material or personal injury, or alleged to be caused directly or indirectly by the information contained in this book.

This book is published with the
cooperation of the
Carrier Corporation
world leaders in the hvac/r industry.

Table of Contents

Chapter **1**
Basic Electricity

This chapter covers many topics concerning electricity, including circuits, wiring, fuses, switches, transformers, drawings of schematics, multimeters and ammeters. This enables a student to better understand what is required to electrically operate and repair air conditioning and furnace controls.

ELECTRICITY

The first step is to understand what electricity is and how it works. Electricity is defined as a form of energy caused by the presence and motion of electrons, protons, and other charged particles. These manifest themselves as attraction, repulsion, electric current, and luminous and heating effects.

One basic rule governs energy: it cannot be created or destroyed. However, it can be transformed into other forms of energy such as heat and light, and can be used as a power source for electric motors.

MATTER

To understand how energy works, it is necessary to understand the construction of matter. Matter is any substance that occupies space and has weight.

Matter is composed of various particles. When matter is broken down into units, these units are called molecules. Molecules, in turn, are made up of atoms, the smallest particle to which an element can be reduced. To clarify, one atom is the smallest part of an element and one molecule is the smallest part of a substance or material. This means that one molecule can either be one atom of an element or else a combination of atoms from two or more different elements. To demonstrate, the water molecule, Figure 1-1 is composed of three atoms. The two smaller atoms represent hydrogen while the large one

represents oxygen. Therefore, a water molecule consists of two atoms of hydrogen (H) and one atom of oxygen (O). The chemical formula for water is H_2O.

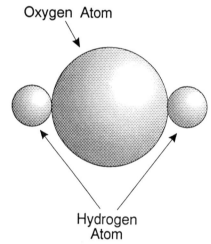

Figure 1-1. Water Molecule

ATOMS

An atom is composed of three elementary particles: electrons, protons and neutrons. These are the three basic building blocks which make up all atoms and, therefore, all matter. Electrons carry a negative charge, protons carry a positive charge, and neutrons do not carry a positive or a negative charge. The significance of these charges will be discussed later.

Figure 1-2 illustrates how electrons, protons and neutrons combine to form an atom. This particular figure shows a helium atom. The helium atom consists of two protons and two neutrons. These protons and neutrons bunch together near the center of the atom and compose the nucleus. Depending on the type of atom, the nucleus contains from 1 to about 100 protons. In all atoms except hydrogen, the nucleus also contains neutrons. The protons and neutrons are similar in weight and size, and their combined weight determines the overall weight of the atom.

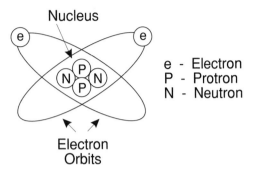

Figure 1-2. *Helium Atom*

The electrons rotate around the nucleus. The helium atom, shown in Figure 1-2, has two electrons. The electrons are extremely light, and they travel at fantastic speeds. The structure of an atom can be compared to the solar system, with the nucleus representing the sun and the electrons revolving around it like planets.

Atoms are too small to see; consequently, assumption, rather than actual observation, is the basis for drawings of atoms. Figure 1-2 represents a very simple drawing of the helium atom based on these assumptions. As of late, more complex models, called Bohr models, of the atom have been proposed. However, all models have this in common: they assume that the basic structure of an atom is that of electrons orbiting around a nucleus, which is composed largely of protons and neutrons.

Figure 1-3 shows the Bohr model of hydrogen. Hydrogen is the simplest atom known. It consists of a single electron orbiting a nucleus composed of a single proton. Hydrogen is the only atom that does not contain a neutron. Because of the simple structure of its atom, hydrogen is also the lightest of all elements.

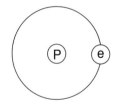

Figure 1-3. *Bohr Model of Hydrogen*

Figure 1-4 shows the Bohr model of the carbon atom. It contains six electrons orbiting a nucleus of six protons and six neutrons. In addition, Figure 1-5 shows the copper atom. It contains 29 electrons, and its nucleus is composed of 29 protons and 35 neutrons. Although not shown, one of the most complex atoms found in nature is the uranium

atom. A uranium atom consists of 92 electrons, 92 protons, and 146 neutrons. The difference between the various elements is that each is made up of atoms which contain a unique number of electrons, protons and neutrons.

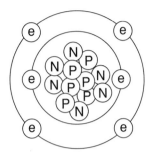

Figure 1-4. *Bohr Model of Carbon*

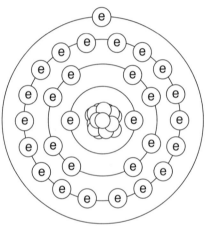

Figure 1-5. *Bohr Model of Copper*

In the examples shown, the number of electrons is equal to the number of protons. This is normally true of any atom. When this is the case, the atom is said to be in its normal, balanced, or neutral state. This state can be upset by an external force. For the most part, though, atoms contain an equal number of electrons and protons.

An electrical charge is a property associated with the electron and proton. The electrical charge is difficult to visualize because it is not an entity, like a molecule or an atom. Rather, it is a property that causes the proton and electron to behave in certain predictable ways.

As mentioned previously, there are two distinct types of electrical charges: positive and negative. The electrical charge associated with the electron is negative, and the electrical charge associated with the proton is positive. The neutron has no electrical charge at all. As its name implies, the neutron is electrically neutral and therefore plays no known role in electricity.

The electron revolves around the nucleus of the atom in much the same way that the earth orbits the sun, in that it is the gravitational attraction of the sun that prevents the earth from flying off into space. The gravitational attraction of the sun exactly balances the centrifugal force of each planet, causing them to travel in more or less circular paths around the sun. The electron revolving around the nucleus is also comparable to the action of a ball which is attached to the end of a string and twirled in a circle. If the string breaks, the ball will fly off in a straight line. It is the restraining action of the string that holds the path of the ball to a circle.

The electron moves around the nucleus at a fantastic speed, so what force keeps the electron from flying off into space? It is not gravity, because the gravitational force the nucleus exerts is much too weak. Instead, the force at work here is caused by the charge of the electron in orbit and the charge of the proton in the nucleus. The negative charge of the electron is attracted by the positive charge of the proton. This force of attraction is called an electrostatic force. Every charged particle is assumed to be surrounded by an electrostatic field which extends for a distance outside the particle itself. It is the interaction of these fields that causes the electron and proton to attract each other.

Figure 1-6 shows a diagram of a proton. The plus sign represents the positive electrical charge. The arrows extending outward represent the lines of force that make up the electrostatic field. Notice that the lines are arbitrarily assumed to extend outward, away from the positive charge. Compare this to the electron shown in Figure 1-7. The minus sign represents the negative charge, and the arrows pointing inward represent the direction lines of the electrostatic field.

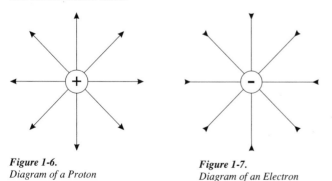

Figure 1-6.
Diagram of a Proton

Figure 1-7.
Diagram of an Electron

ELECTRICAL CHARGES

There is a basic law of nature which describes the action of electrical charges: like charges repel, and unlike charges attract. Because like charges repel, two electrons repel

each other as do two protons. Figure 1-8 illustrates how the lines of force interact between two electrons. The directions of the lines of force are such that the two fields cannot interconnect. The result is that the electrons attempt to move apart, or repel each other. The same repulsion occurs with two protons, as shown in Figure 1-9. However, in the case of an electron and a proton, Figure 1-10, the two fields do interconnect. The two opposite charges attract and tend to move together.

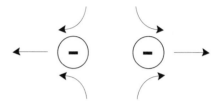

Figure 1-8. *Electrons Repel*

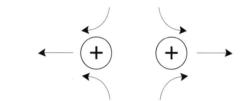

Figure 1-9. *Protons Repel*

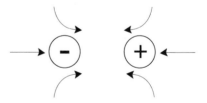

Figure 1-10. *Electron and Proton Attract*

ELECTRON FLOW

Atoms tend to remain stable, that is, they contain a constant number of electrons and protons, so an additional force is necessary to move the electrons from one atom to another in order to create electrical flow. Electrons move from negative to positive. If a deficiency of electrons is created—if the atom has fewer electrons than it is supposed to have—then the electron composition is unequal. As a result, a potential difference of electrostatic force exists. This difference is called electromotive force (emf) or voltage (V). This force pushes the electrons from one atom to another, resulting in electron flow, Figure 1-11. Some electrons can move from one atom to the next, because their attraction to the nucleus is weak. These electrons are called free electrons. These free electrons are most easily moved from one atom to the next by the emf.

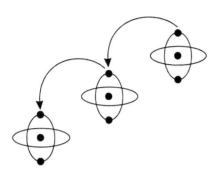

***Figure 1-11.** EMF Creating Electron Flow or Current*

Electrons travel at a speed of 186,000 miles/second, which is the speed of light. While this rate remains constant, the number or density of the electrons passing a given point at a given instant can change. Thus, a greater or lesser current flow results. The quantity of electron flow, or current, is measured in amperes (A).

All materials resist the flow of electrons to a certain extent. The extent of this resistance to flow, which is measured in ohms, determines whether the substance is considered a conductor or an insulator. Glass, plastics and mica have a high resistance to electron flow and are often used as insulators, Figure 1-12. Many metals such as copper, silver, aluminum and platinum have an excess of electrons, so the electrons pass freely. These materials are normally used as conductors, as it takes relatively little force to start and maintain electron flow.

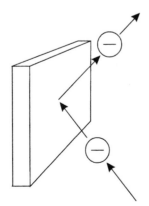

***Figure 1-12.** Insulators Resisting Electron Flow*

Even with good conductors, some other factors will increase resistance to electron flow; ambient temperature (the temperature around the electrical conductor, or wire)

is an important factor. When the surrounding area is cold, there is very little resistance. However, if the wire is confined in a small area where the temperature can build up, or the wire is operating in a hot surrounding atmosphere, resistance to current flow increases. Wires are insulated not only from each other, but also to reduce the effect of ambient temperature.

The area or size of the wire can also affect flow. If the cross-sectional area of the wire is too small for the current carried, friction causes the wire to warm up and reduce the flow of electrons. Excessive heat can also build up and burn the insulation, causing a short. No hazard exists if the wire is oversized, because this reduces friction, which in turn, reduces the amount of heat generated.

At one time or another everyone has seen or felt the effects of electrostatic charges. One of the most spectacular displays of an electrostatic charge is lightning. Less spectacular examples include: receiving a mild shock when removing clothes from a dryer, combing hair, or touching a metal object after shuffling along on a rug. In each of these cases, two different bodies receive opposite electrical charges. This happens when one of the bodies gives up a large number of electrons to the other. The body that gives up the electrons becomes positively charged, while the body receiving the electrons becomes negatively charged.

Light, heat, magnetism, pressure and chemical activity all produce an electromagnetic force. It is interesting to note that the reverse is also true. That is, an emf can produce light, heat, magnetism, pressure, and chemical activity. An example of electricity producing light is the light bulb. Examples of electricity producing heat include the toaster and electric stove. When current flows through a wire, a magnetic field surrounds the wire. This magnetic field is put to practical use in motors, loud speakers and solenoids.

ALTERNATING CURRENT

The power companies generate what is known as alternating current (ac), or current flowing first in one direction and then another. This is the most common current with which a service technician works. The voltage encountered is a 120-V, single-phase, 60 cycle (also referred to as Hertz) alternating current. Figure 1-13 illustrates how the current flow changes direction 60 times/second. The voltage builds to a positive peak (90 degrees) from Point A to Point B, then diminishes to zero at Point C. It then declines to a negative or minus value at Point D and returns to zero at Point E.

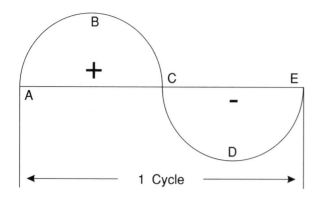

Figure 1-13. *Alternating Current*

DIRECT CURRENT

Direct current (dc) travels in one direction and flows from negative to positive. This is because electrons, with a negative charge, travel to atoms carrying a positive charge.

WIRING DIAGRAMS

Wiring diagrams denote one wire (H) for hot, and this wire is black. The other wire is marked (N) for neutral, and this wire is white. Single-phase, 120-V current requires only these two wires. The neutral wire is sometimes called a ground wire; this should not be confused with an earth ground wire. The neutral wire is actually the neutral or zero voltage side of the circuit. A third wire, called an earth ground, is added to certain circuits. The earth ground wire is green. Most installations include only the ground (neutral) wire in the outlet, which will accommodate a three prong plug. On wiring diagrams, the earth ground wire (third wire) is not usually shown.

CIRCUITS

An electrical circuit must have a power source, a continuous path of conductors to carry the current, and some sort of device to use the current. There are three basic types of circuits: series, parallel, and series-parallel.

SERIES CIRCUITS

A series circuit is a one-path circuit; that is, it never has more than one conductor connected to one terminal and there is only one path from the source to the load, then back to the source. In a series circuit, the loads are connected end-to-end. Using batteries as an illustration, the series circuit is described below.

In a 12-V automobile battery, six cells are connected so that the individual cell voltages add together. In the 6-V battery, three cells are connected in the same way. This

arrangement, called a series connection, is shown in Figure 1-14 as it occurs in a 3-cell flashlight. The cells are connected so that the positive terminal of the first cell connects to the negative terminal of the second; the positive terminal of the second connects to the negative terminal of the third, and so on. This is a series connection because the same current flows through all three cells. Since the individual emf of each cell is 1.5 V, the overall emf is 4.5 V. The schematic diagram for this connection is also shown in Figure 1-14.

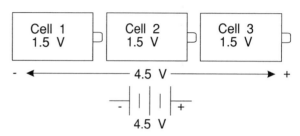

Figure 1-14. *Series Connection in 3-Cell Flashlight*

Figure 1-15 shows a different type of cell. Here again, the three cells are wired together in series. Notice that the voltages add together because between cells the opposite polarity terminals are connected. That is, the negative terminal of the first cell connects to the positive terminal of the next, and so on. Thus, the three 1.5-V cells provide a total emf of 4.5 V.

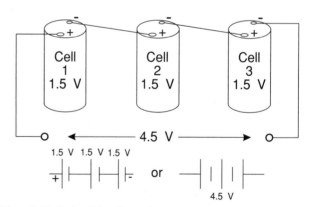

Figure 1-15. *Series Aiding Connection*

With the series connection, the total voltage across the battery is equal to the sum of the individual values of each cell. However, the current capacity of the battery does not increase. Since the total circuit current flows through each cell, the current capacity is the same as for one cell.

Light bulbs with voltage starting at 120 V are represented in Figure 1-16. As the voltage travels through each resistance, there is a drop in voltage proportionate to the amount of resistance. Upon return to neutral, the voltage is zero. With a 120-V resistance, there is a drop of approximately 20 V across each resistance. If only three bulbs are used, each bulb glows more brightly, with each having about twice the voltage drop (40 V) to produce more light. In a continuous circuit, it does not matter which side of the resistance is wired to (H) and which continues to neutral.

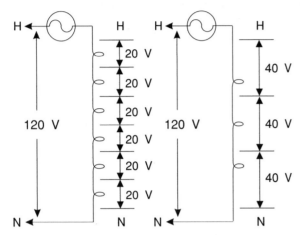

Figure 1-16. *120-V Circuit with Various Resistances*

As shown in Figure 1-17, current passes through each resistance in sequence - from the first, to the second, to the third, to the fourth, and so on. As there is no alternate path, when one bulb is unscrewed, it acts like a switch and the entire circuit de-energizes. This is why several safety devices, controlling the same component, are wired in series with the component. If any one of them senses trouble, it takes the component off the line. Switches are always wired in series and ahead of the device they control.

When there are several unequal resistances in a series circuit, then the voltage drop is greatest across the highest resistance. A 100-Watt (W) bulb has larger wire than a 40-W bulb, so the resistance of the 100-W bulb is much less than the 40-W bulb. If both are wired in series, most of the voltage is used to light the smaller bulb and the larger one does not have enough voltage travel through it to light.

The bulbs glow brightest when the circuit is first energized. The bulbs then dim somewhat after a short period of time. The reason for this is, as the bulbs warm up, they offer more resistance, causing less current flow. As a result, they do not glow as brightly. The current in a series circuit is the same at all points in the circuit, regardless of the number and amount of resistances.

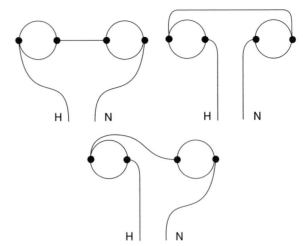

Figure 1-17. *Alternate Methods of Wiring in a Series*

PARALLEL CIRCUITS

As mentioned earlier, the series connection of cells increases the output voltage but not the current capabilities of the cells. However, there is a way to connect cells so that their current capabilities add together. This is called a parallel connection and is shown in Figure 1-18. Here, the loads are thought of as being side-by-side (parallel), rather than end-to-end. A parallel circuit has more than one path between the two terminals of the source of power.

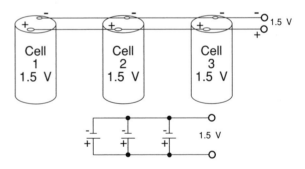

Figure 1-18. *Parallel Connection*

The total resistance of the parallel circuit is always less than the smallest individual resistance of the components. Current (amps) divides equally among the branches, should the resistances in a parallel circuit be equal. For example, if two bulbs, each with a resistance (R) of about 1200 ohms, are wired in parallel, the total resistance (Rt) in the circuit is 600 ohms.

$$\frac{1}{Rt} = \frac{1}{1200} + \frac{1}{1200} = \frac{2}{1200} = \frac{1}{600}$$

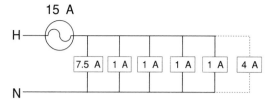

Voltage divided by resistance equals the total current. In this case, 120 V divided by 600 ohms equals 0.2 A, and each of the parallel circuits draws 0.1 A. Should the resistances differ, the amps always divide proportionately.

Using the bulb example, each bulb in a parallel circuit burns with full intensity; they do not dim as they did in a series circuit. This remains true even if one circuit has a 100-W bulb and another a 20-W bulb, Figures 1-19 and 1-20. For this reason, house lighting circuits are wired in parallel; each circuit then has full voltage and the bulbs all light equally.

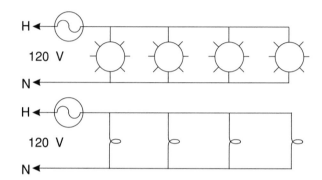

Figure 1-19. Pictorial and Schematic Diagrams of a Parallel Circuit

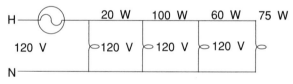

Figure 1-20. All Loads in a Parallel Circuit Have Equal Voltage

The total amperage in the circuit is the sum of amps drawn by all the branch circuits. It is for this reason that the addition of another appliance to a circuit already near capacity causes a fuse to blow. Take, for example, an air conditioner that draws 7-1/2 A from a circuit. Also on this circuit are four other branches, each drawing 1 A, for a total load of 11-1/2 A, Figure 1-21. The circuit is fused for 15 A, and even if all loads are energized, there is no problem. The problem occurs, however, when adding another appliance. This extra appliance draws over 3-1/2 A, overloads the circuit, and finally blows a fuse.

Major components in the system, such as blower motors, use parallel circuits, because the voltage drop remains constant and is the same as the supply voltage.

Figure 1-21. Total Amperage in a Parallel Circuit is the Sum of the Amps Drawn By All the Branch Circuits

SERIES-PARALLEL CIRCUITS

In an actual installation, there are combinations of series and parallel circuits. Some series circuits are in parallel with each other, and a series circuit may include components wired in parallel with each other. These circuits are called series-parallel or parallel-series.

The cells are connected in series-parallel when both a higher voltage and an increased current capacity are required. For example, when four 1.5-V cells need to be connected so that the emf is 3 V, and the current capacity is twice that of any one cell. This is achieved by connecting the four cells as shown in Figure 1-22. To achieve 3 V, cells 1 and 2 are connected in series. However, this does not increase the current capacity. To double the current capacity, a second series string (cells 3 and 4) must be connected in parallel with the first. The result is the series-parallel arrangement shown.

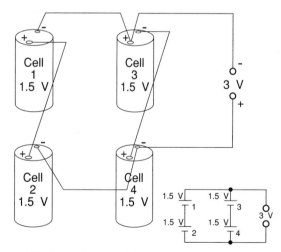

Figure 1-22. Connecting Four Cells

Whether a circuit is in series or parallel depends on its electrical relationship with another circuit or component. In Figure 1-23, circuit AC is a series circuit wired in parallel to series circuit BC. Both are in parallel to circuits DF and EG. To be part of a series circuit, a component may be wired in parallel with another component.

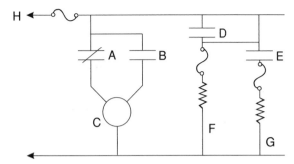

Figure 1-23. Parallel-Series

CIRCUIT PROTECTION DEVICES

All electrical circuits must have protection devices to guard against current overloads. When too much current flows through the circuit, the wires and other components overheat, causing damage. Fuses, circuit breakers, and switches all serve as circuit protection devices.

FUSES

Fuses protect wiring and other components of a circuit from burning up should overheating, a short, or overcurrent occur. Most fuses contain a length of metal which has a low melting point and a higher resistance than the circuit conductors. When the current exceeds a predetermined point, the metal melts and opens the circuit.

There are many different types and varieties of fuses for different types of applications. A circuit can have a single fuse, or a main fuse and several branch fuses. A 120-V circuit must have at least one fuse in the hot leg (H), and it should be located as closely as possible to the beginning point.

The terms short and overcurrent have very different meanings. Overcurrent occurs when amps are added to a full load. For example, when an addition to the circuit causes an increase from 15 to 17 A. After several minutes, the fuse blows. Specifically, an overcurrent is when a fuse element melts, arcs, vaporizes and finally opens, causing an open circuit.

When a short occurs, the circuit immediately overheats the wire, causing the fuse to blow at that instant. Due to a lack of resistance in a short, the load also increases immediately. The flow of electrons increases the friction in the wire and heats it up rapidly. This makes it necessary to stop that flow before the insulation on the wire burns up.

In the case of a short circuit, the entire center of the fuse heats instantly, and the extreme temperature causes it to vaporize. This vapor, in turn, carries current and supports an arc, which fills the case with an arc-extinguishing gas

that cools and condenses the vapor. This stops the arc and breaks the circuit.

Types of Fuses. Various circuit applications require different types of fuses, Figure 1-24. For lights, electric heaters, some appliances, and circuits with only resistance loads, the simple strip fuse screw in a breaker bar type housing can be used. These come in many amperage ratings, but they must be sized to the circuit load.

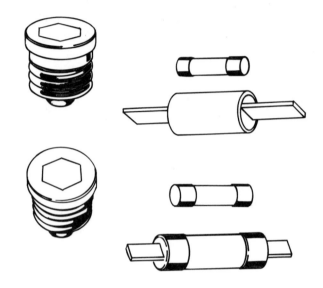

Figure 1-24. Various Types of Fuses

A circuit containing motors draws five to seven times the amperage on startup, so it needs a fuse large enough for the starting load to protect against lesser overcurrent while it is running. At the same time, the fuse cannot be too small or it will constantly blow on start-up. To solve this problem, a dual element fuse, called a time delay fuse, is used, Figure 1-25. One part of the fuse is a metal strip that is designed to carry higher currents for a short time, while still protecting against a short. A motor, when operating normally, comes up to speed quickly. Therefore, a short delay before blowing the fuse gives adequate protection. A second element, a solder pot, is used for overload protection. When current overload continues, the solder melts and releases a spring that opens the circuit. This is especially useful when the overload continues beyond the time required for the motor to come up to speed.

Replacing a Fuse. Whenever a fuse blows, the trouble or cause must be found. When replacing a fuse, the disconnect must be opened first. The fuse should not be inserted in a hot circuit, as this could cause an arc, resulting in a burr or rough spot on the blade or cap. This then causes poor contact. In order to enable a good contact with the fuse and the fuse holder, the fuse holder should always be

Fusible Strip

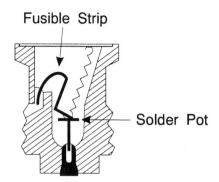

Solder Pot

Figure 1-25. Time Delay Fuse

checked to make sure it is clean and rust-free. A bad connection can cause a percentage of its rated amperage to drop, and additional current causes overheating. An easy way to check for a bad connection is to insert a narrow piece of paper between the clips and the fuse cap or blade. If the paper passes through at all, it is a bad contact, Figures 1-26 and 1-27.

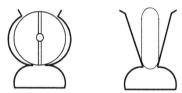

Figure 1-26. Poor Contact with Fuse Holder

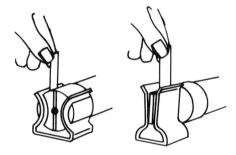

Figure 1-27. Check Fuse Contact with Holder

In the booklet *Fuseology*, Bussman Manufacturing Company states that whenever a fuse blows, it is due to one of the following reasons:

- Short Circuit: The cause of the short circuit should be checked and corrected before a new fuse is inserted.
- Overloaded Circuit: Check the total circuit amperage draw. If the amperage is higher than the manufacturer specifications, the problem must be found and corrected before inserting an additional fuse.
- Poor Contact: Vibration, ambient heat, or a bent fuse holder can all cause poor contact. Discoloration of the contact surfaces, either on the fuse holder or the fuse itself, frequently indicates poor contact. To ensure positive contact, the fuse holder must hold the fuse

firmly and in proper position. The fuse must be in contact with both sides of a spring-type clip. If the clip is bent, as shown in Figure 1-28, reforming the fuse holder ensures proper contact. Any improper contact or loose connection generates increased heat; this can cause the fuse to blow. In Figure 1-29, lug A must be properly soldered to the wire or cable; bolt B, holding the lug to the fuse holder, must be tightly drawn up; and screw C, holding the clip to fuse holder, must hold the clip firmly. Switch blade clips F and switch hinge G must be tight.

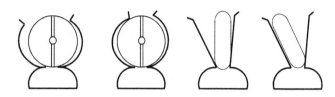

Figure 1-28. Reforming Bent Fuse Holders

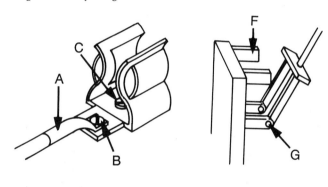

Figure 1-29. Secure Connections

- Wrong Size Fuse: If the fuse is too small, it blows immediately, as it cannot handle the load on the circuit. If the fuse is too large, it does not protect the circuit and blows only when a serious problem occurs. Following the manufacturer's instructions and using the proper size fuse alleviates these problems.
- High Ambient Temperature: If the air around the fuse is very hot — for instance, if the fuse is in an enclosed box with the heater or some other source of heat — it has a greater chance of blowing. Maximum temperature is approximately 125 °F. Should the temperature rise above this point, inadequate protection results.
- Vibration: Vibration can cause the fuse holder to change its shape; subsequently, the fuse holder cannot make proper contact with the fuse. When this happens, additional heat is generated and the fuse may blow, even before the current becomes excessive.

CIRCUIT BREAKERS

Because fuses are designed to give one-time protection, they must be replaced if they blow. However, one type of circuit protection has a manual reset feature called a circuit breaker, Figure 1-30, and this serves as the main protection at the service entrance for all branch circuits. The same principles of a fuse apply to the circuit breaker, in that the circuit breaker senses heat from overcurrent and trips a switch when the heat becomes excessive. Some small fuses, usually found on controls or instrument panels, are available which are manually reset.

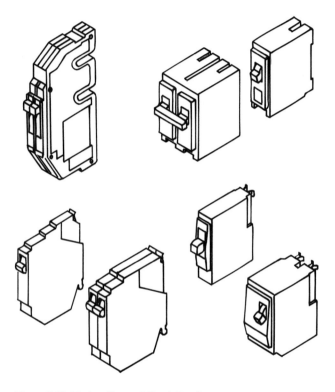

Figure 1-30. *Various Types of Circuit Breakers*

SWITCHES

The operation of all electrical devices in the circuit is controlled by a switch, which is located in the hot line ahead of the devices being controlled. These switches are classified by the number of poles. For example, a single-pole switch has one hot line; a double-pole switch has two hot lines. Switches are also classified by the number of throws; for example, the single-pole, single-throw switch is comparable to a simple on-off switch, controlling one load with one load, one source. Another example is a switch on a house lighting circuit. Even though several lights are on the same circuit, the switch supplies power to only one circuit. A single- pole, single-throw switch is identified as SPST.

A single-pole, double-throw (SPDT) switch controls alternate loads which have two loads and one source. An example is a heating-cooling switch on a thermostat. It has one source and two loads. It may also have an off position, but this is not counted as a load.

A double-pole, single-throw (DPST) switch consists of two hot poles and only one load. For example, an oven contains two 120-V power sources which combine to run one 240-V appliance. Two power sources and two loads identify a double-pole, double-throw (DPDT) switch.

A rotary switch may have one hot pole and many choices of loads, or a single-pole switch with multiple throws or loads. A slide switch can consist of two single-pole switches mounted together with an off position. In either position, only one source and one load are connected.

Switches operate and control the circuit as well as serve as safety devices. To ensure safe replacement, they should always open on the power side of the fuse. The switch should always be in the hot leg ahead of the device it is supposed to control. This is a safety precaution, in case there is a short to ground. With the switch ahead of the load, the load cannot operate with the switch open. If the switch is closed and there is a direct short, the fuse blows, stopping all power. If a short to ground occurs between the load and the open switch in the neutral leg, the load still operates, the fuse does not blow, but a serious shock hazard exists. Switches in the heating and air conditioning systems may operate on either low or high voltage.

A switch must perform a mechanical action in order to work; in other words, a switch must open or close a set of contacts. Therefore, a switch is considered to be normally open or normally closed. If a switch is normally open (NO), this means it is open unless activated. The other position is normally closed (NC), meaning the switch stays closed unless activated, Figure 1-31.

Figure 1-31. *Normally-Open and Normally-Closed Switches*

Manual Switch. The simplest and most frequently used switch is called a manual switch. A furnace disconnect or a fan switch at the thermostat may contain this type of switch. As the name suggests, the mechanical action (opening or closing the switch) is a manual operation, and the switch remains in the desired position until it is manually moved again. A manual switch can accomplish any of the previously described actions (SPST, SPDT, etc.). The switch symbol on a diagram is shown as a dashed line through the switch (or switches) to indicate that it is a manual switch, Figure 1-32. For safety reasons, a manual switch always breaks the circuit at the line contact end of the switch.

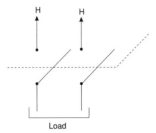

Figure 1-32. Manual Switch

A magnetic coil, constructed of wire wound around an iron plunger which is free to move up and down within the wire coil, Figure 1-33, operates manual switches. The switches may be either relays or solenoids. Low voltage usually energizes the coil of wire. When this occurs, a magnetic field is set up which pulls the plunger up into the coil. Relays are equipped with electrical contacts at the end of the plunger which make (close) or break (open) as the plunger moves upward, switching the second circuit which usually operates on line voltage.

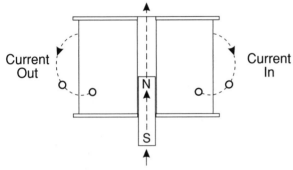

Figure 1-33. Magnetic Coil

When the coil is energized in a normally-closed (NC) switch, the plunger moves up, opening the electrical circuit. The upward movement of the plunger closes the circuit in a normally-open (NO) switch.

Heat or Temperature Switch. When switching by heat or temperature, a bimetallic element is commonly used, such as in a thermostat. A bimetallic element is composed of two dissimilar metals bonded together. Because the two different metals react differently to temperature changes, the warping action of the element can be used to activate switches. The bimetal changes position when the ambient temperature varies due to heating or cooling. By placing an electrical contact on the end, the contact can open or close due to this action. A heat-actuated switch can be NO or NC and be operated by this rise or fall in temperature; this is usually a single-pole, single-throw switch, Figure 1-34.

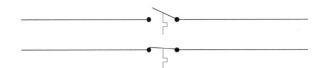

Figure 1-34. Thermostat with Heat-Sensitive Bimetallic Switch

Pressure Switch. Pressure is another method used to open or close a switch. Pressure is directly related to temperature in a confined system. As such, pressure controls are set to correspond to safe operating temperatures. Pressure switches are commonly found on air conditioning systems, as compressors can be permanently damaged if operating pressures exceed or fall below the normal operating pressure.

A combination high-low (dual) pressure control, Figure 1-35 is a very common pressure control. The pressure in the system acts as a bellows which expands or contracts against a preset spring pressure. Pressure overcomes the spring pressure, tripping a mechanical linkage which then opens a set of electrical contacts. Some units have an individual high and low pressure control which senses temperature and trips when the temperature reaches the corresponding pressure trip point. The sensing probes are soldered directly into the suction and discharge lines.

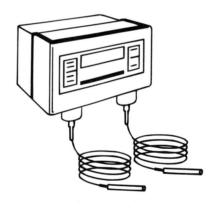

Figure 1-35. High-Low Pressure Control

Light Switch. Light is another basic method used to actuate a switch. Positioned so it can see the pilot and main burner at all times, the cad cell, also called a fire eye, requires precise alignment at exactly the right angle to the pilot and main burner so it can actuate protective controls on the fuel feed in case the pilot light is extinguished.

The main monitoring component of a typical cad cell is a lead sulfide cell that is approximately the size of a small glass marble. The sulfide cell has the unique capability of conducting electricity when it detects light in the visible range of the spectrum. When the cell detects light, it conducts enough electricity to transmit signals to a sensitive relay in the primary control. The transmission indicates that ignition has occurred and fuel can continue to flow to the burner. In the dark, the cell does not conduct electricity. Consequently, the primary control does not allow fuel to flow to the burner.

The cad cell, Figure 1-36 supervises the flame throughout its entire burning cycle. If the fire goes out for any reason, the electrical transmission from the cad cell stops and the fuel valve is automatically shut off by the primary control. A purge cycle is then initiated to clear the furnace or fire box of all combustible gases before pilot ignition is attempted and the cycle restarted.

Figure 1-36. Cad Cell

Placement of the cell is critical, Figure 1-37. The cell must see light from the flame but it cannot mistake transient light from other sources, such as an open inspection door, for a flame. On installation, the cell is carefully positioned by the manufacturer and its location should not be disturbed.

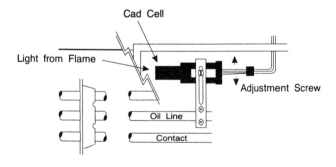

Figure 1-37. Proper Position of Cad Cell

Alternate types of flame detectors are sometimes used, but the majority of boilers in operation today utilize a fire-eye sensory scanner.

Moisture Switch. Moisture is a specialized method of opening or closing a switch. Certain substances, such as wood or human hair, swell with the addition of moisture, and shrink as the moisture content reduces. This principle applies when operating a humidifier. A component called a humidistat senses the moisture content in the air with its element and opens or closes a switch, thus activating a humidifier.

In order to control an electrical circuit and produce the desired results, all of these various methods of opening and closing a switch can be used, alone or in combination. At this point, it is important to remember that all circuits contain switches of some sort and that some circuits may have several switches controlling the same device.

Transformers. A transformer has no moving parts, and its job is to change one ac voltage to a different ac voltage. The transformer consists of a steel core with two coils or windings, one on each side of the core, Figure 1-38. These windings are insulated from each other, as well as from the core.   The line side, called the primary, is the side connected to the supply voltage. The load side, called the secondary, delivers the energy at a different voltage, Figure 1-39.

Figure 1-38. Transformer. Courtesy, Honeywell Incorporated.

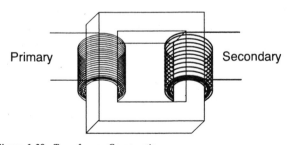

Figure 1-39. Transformer Construction

Most of the safety devices used in air conditioning and heating operate on low voltage circuits; 24 V rather than line voltage. The reasons for this are:

- Wiring for low voltage is much less expensive to run because wire size is smaller and easier to handle.
- Low voltage wiring does not require conduit. This keeps cost down and presents little, if any, fire or shock hazard.
- Switches, coils, and other components designed for low voltage are less expensive than similar components designed for line voltage.

A transformer is called a step-down transformer if 120 V is supplied to the primary and the secondary delivers 24 V. A step-up transformer is used to increase the voltage. An example of a step-up transformer is 120 V on the line side producing 240 V on the load side.

The primary is always the line side, but is not necessarily the highest voltage. It is important to remember that the wire turns determine the voltage. For example, stepping down from 120 V to 24 V has five times the number of wire turns on the primary as on the secondary. The current amps increase by five times; 120 V with 1 A becomes 24 V with 5 A, Figure 1-40.

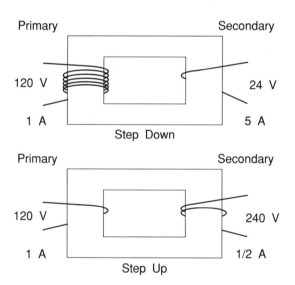

Figure 1-40. Step-Up and Step-Down Transformers

The power output of the secondary winding, in terms of volt-amps (voltage multiplied by amps, or vA), determines how transformers are rated. Ambient heat, applied load and internal voltage drop can all affect this rating. However, even though the load on the secondary changes, the voltage remains steady.

Correctly-sized transformers are a must, and they also must have enough power to handle the maximum load in the circuit. There is a vA rating for each circuit, and this is the minimum size transformer which can be applied. For example, if a 24-V control circuit draws 1.4 A of

current, the transformer vA needed is 24 x 1.4, or approximately 34 vA. As is true of sizing wire, a larger transformer can be used with no detrimental effect on the circuit.

A fusible line protects the secondary of many transformers against overload. However, if this fusible line blows, it requires the replacement of the entire transformer. Some transformers have replaceable fuses, in which case only the fuse needs to be changed. Primaries are protected by the fuses in the line.

CONTROL CIRCUITS AND DIAGRAMS

As stated previously, 24-V systems use control circuits for the following reasons: lower cost of wiring and components, and safety shock hazards. Commercial equipment and electric heat applications usually use line voltage control circuits. These are generally shown with a dotted line or light line to distinguish them from line voltage circuits. Normally, low voltage circuits use a relay coil to open or close a set of line voltage contacts. The coil used in low voltage is in series with the actuating device, thermostat, limit switch, etc. The line voltage contacts are in series with the device to be controlled, e.g., a compressor blower.

There are two basic methods of drawing a circuit diagram: (1) pictorial wiring diagrams, and (2) schematic wiring diagrams.

PICTORIAL WIRING DIAGRAM

One method is to draw the diagram with the components and wiring in about the same relationship as is the actual installation. This is commonly called a pictorial wiring diagram. When in the field, this type of drawing is helpful in determining the actual wiring equipment needed, as well as the number of wires required for the job. Figure 1-41 shows a common heat only circuit.

The pictorial diagram is very useful in cases of simple wiring. However, as the wiring becomes more complicated, routing and identifying individual wires becomes very difficult. From a service standpoint, a pictorial wiring diagram in this situation does not adequately show the relationship of some of the components, or which component is in which circuit.

SCHEMATIC WIRING DIAGRAM

It is for the reasons listed above that there is another method for complicated wiring, called a schematic wiring diagram. This type is the most commonly used. When using this method, the physical wiring is disregarded, and the diagram itself becomes totally functional. Symbols represent the components in the schematic diagram. Consequently, a constant review of symbols is necessary in order to learn the symbols by memory. In a schematic wiring diagram, the line voltage circuit is shown as two horizontal, or parallel, lines. A solid line denotes line voltage, and a dotted line denotes low voltage, Figure 1-42. It is interesting to note that a relay may have its contacts in the line voltage circuit, while the coil is

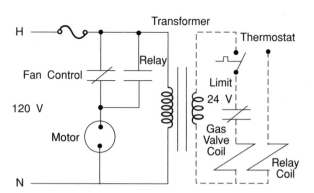

Figure 1-42. Schematic of the Gas Furnace Control Circuit

In drawing a schematic, it is necessary to first identify components that are in the same circuit, and whether or not succeeding circuits are in parallel or in series. In order to advance further, it is very important to learn about, practice and study all types of schematics.

120-V AND 240-V POWER

As previously discussed, 120-V, single-phase power uses two wires: one hot wire and one neutral wire. When two, 120-V, single-phase power sources are put together with a common ground, a 240 V is the result. This is a three-wire service with the neutral wire grounded at the source, two hot wires running to the load, and the third wire grounded at the disconnect.

If a reading is taken from each of the two hot wires to the neutral, it will read 120 V. A reading across the two hot wires will read 240 V. Larger motors and heavier equipment commonly use this type of service, and they are still alternating current just as in the 120-V service, Figure 1-43.

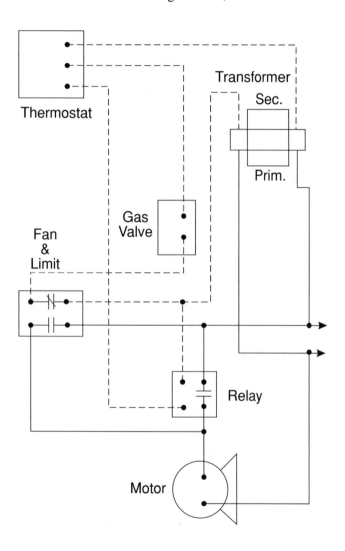

Figure 1-41. Pictorial Drawing of Gas Furnace Controls

shown some distance away in the low voltage circuit. Even though the same housing encloses these two elements, separating the line and low voltage circuits makes the drawing much easier to read. This drawing also defines what happens down the line if a given switch is open.

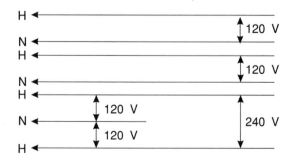

Figure 1-43. Two 120-V Sources Can Be Linked to Provide 240-V Power

MULTIMETERS AND AMMETERS

There are several different types, shapes and sizes of meters, as well as varieties in scales. The meters normally used are multimeters and ammeters.

MULTIMETERS

The term multimeter refers to an instrument having two or more meters, such as volts, ohms, and amps combined into one meter. Multimeters are also referred to as volt-ohmmeters, Figure 1-44.

Figure 1-44. Multimeters. Courtesy, A. W. Sperry Instruments, Incorporated.

All volt-ohmmeters have two leads, one red and one black. The red lead is the positive (+) lead, and there is a jack on the front of the meter marked (+), or positive. The black lead is the negative (-), or common, lead. The leads have an insulated shoulder which should be used to insert or remove the lead from the jack. It is necessary to not pull the leads out by the wire; instead, the insulated shoulder should be held. Some probes have alligator clips on the ends of both leads, while others may have an alligator clip on the common lead and a straight metal probe on the plus (+) lead. Other leads may have a metal probe.

Most meters have a main range switch for selecting the desired circuit: resistance, volts alternating current (vac), volts direct current (vdc), and sometimes milliamperes. In turn, each circuit has as many as six ranges.

Some meters have the scales only, and a separate knob selects alternating current (ac) or direct current (dc). While other voltage scales can be used, the most common voltage scales used occur in multiples of five (10, 50, 250, and 500 V) or multiples of three (0.6, 3, 60, 300, 600, and 3000 V). With some meters, the probes must be moved to separate jacks to select the very low or very high voltages. Meters with scales in multiples of three are not recommended for heating and air conditioning work.

Since voltages often run over 120 V (122 to 126 V), using the 120-V scale in this situation burns out the meter. It is also difficult to get an accurate reading of 122 V on the 300-V scale. There is a minimum of five scales on the meter face: two for ac, two for dc and one for ohms. The ac scale is usually red; all others are black. These scales are calibrated to match the range selector knob. In the case of a scale and selector set up for multiples of five, the ac scales will be from 0-10 for ranges of 10; 0-250 for ranges of 250; and 0-50 for ranges of 50 and 500, Figure 1-45.

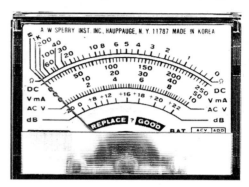

Figure 1-45. Face of Volt-ohmmeter. Courtesy, A. W. Sperry Instruments, Incorporated.

Since high voltage can exist even when not expected, the range that is initially selected should be at least one range higher than the expected voltage. Putting a meter across a voltage higher than the top of the selected range burns out the meter. The probes are placed on the H and N sides of the load. With the selector on the 500-V range, a 120-V source moves the needle between 10 and 20 on the 50 scale, indicating 120 V. When the selector switch is moved to the next lower scale (250 V), the needles swing up to 120 on the 250 scale, again indicating 120 V. The lowest scale encompassing the voltage being measured produces the most accurate readings. In order to give a more accurate reading, many meters have a mirror behind the scale to eliminate parallax. It is always necessary for the service technician to exercise caution when placing the probes in the equipment, particularly to avoid shorting. Touching the probe to any other metal can cause a direct short.

The dc scales are handled in exactly the same manner as the ac scales. The resistance (ohm) circuit usually has three ranges: Rx1, Rx100 and Rx10k. The Rx1 scale reads resistance directly (multiplies by one); the number read on the Rx100 scales must be multiplied by 100 to get the value; and the reading on the Rx10k scale is multiplied by 10,000 to obtain the value.

The resistance scale usually reads from 0 on the right to infinity (no limit) on the left. When not activated, the needles rest on the left-hand side. The meter has a knob labeled ohm adj. This is used to zero-in the scale for the most accurate reading. The following steps should be performed to test the accuracy of the meter:

1. Place the selector in the Rx1 position.
2. Touch the two probes together. The needle should swing all the way to the right and stop at zero.
3. If the needle stops short or goes beyond, turn the adjustment knob until the needle is exactly on zero. Read the ohms for this range.
4. Repeat this adjustment for each range and check before a resistance reading is taken.

If the adjustment knob does not bring the needle to zero, this indicates that the batteries in the meter are weak and should be replaced. As the size and number of batteries varies between meters, the back must be removed in order to select exact replacements.

AMMETERS

For the most part, ammeters are used to check amperage. Ammeters may have lead attachments in order to read voltage and ohms. This is performed in much the same manner as previously described for a volt-ohmmeter. One unique feature of some ammeters is that a knob selects the scale and then positions the selected scale in a window in the handle, Figure 1-46. Only this scale is visible. Others have multiple scales similar to volt-ohmmeters.

Figure 1-46. *Selected Scale in Window of Handle of Ammeter. Courtesy, A. W. Sperry Instruments, Incorporated.*

An ammeter has a set of jaws that can be opened by pressing a trigger or lever on the handle. This allows the jaws to slip over and go around a wire in the circuit, therefore eliminating the need to disconnect the wire. Amperage is read by encircling one wire of the circuit.

Typical scales are 0-6, 0-15, 0-40, and 0-100 A. However, some have 0-300 A. These scales are selected in advance. Again, in order to avoid damage to the meter, the highest scale should be selected.

As amperage readings are sometimes taken in dark or inconvenient areas, many ammeters have a button on the handle that locks the pointer in one position when the reading is taken. Then, the ammeter holds the reading after the probe is removed from the wire. To release the reading, the button is moved to the left; to the right to lock the needle in place. This feature allows measurements to be taken in total darkness. When moved into the light, the meter displays an easy-to-read scale.

Making an Amperage Multiplier. A service technician can make an amperage multiplier. This is an easy way to assure greater accuracy when measuring small amperages. To make an amperage multiplier, the following steps should be performed:

1. Using light wire, make a loop with exactly ten turns (the wire size or the diameter of the loop do not matter).
2. Place alligator clips on each end of the loop and tape the loop in two or three places. This is necessary to hold the wires together.
3. Attach this loop in series to the circuit.
4. Place the jaws of the ammeter around the loop. The loop makes the value registered on the dial ten times greater than the actual current. Using the smallest scale on the ammeter and dividing the reading by ten gives a far more accurate reading.

SAFETY

When working with electricity, safety is the service technician's top priority. The following rules are common sense but are worth repeating:

■ Never depend on another person to shut off the power source—always check the power source yourself. Many accidents occur due to someone turning the wrong power off or simply misunderstanding the unit being serviced.
■ Never stand in water when touching a live wire. This allows current to pass through the body to the ground, which can cause serious injury or fatality.
■ Always treat any circuits as live until you have checked them and are absolutely positive that they are not.

Electricity does not need to be feared if handled properly.

REVIEW QUESTIONS

1. What is a molecule?
2. What does an atom consist of?
3. What are the charges of an electron? proton? neutron?
4. What keeps the electron orbiting the nucleus?
5. What is voltage (V)?
6. Name two factors which increase the resistance to electron flow.
7. What is alternating current? direct current?
8. What is an ampere (A)?
9. Describe a series circuit and how it differs from a parallel circuit.
10. What is a series-parallel circuit?
11. Name two circuit protection devices and how they function.
12. What is a short?
13. What is overcurrent?
14. Describe a time delay fuse.
15. State three reasons why a fuse might blow. Describe each reason.
16. Where are switches located?
17. How does a manual switch work?
18. How does a pressure switch work?
19. What is a cad cell?
20. What is the purpose of a transformer?
21. Describe the two methods of drawing circuit diagrams.
22. What is a multimeter? What does it measure?
23. What is an ammeter? What does it measure?
24. Name the three major safety rules a service technician should follow when working with electricity.

The most highly engineered precision component in the heating and air conditioning system is possibly the thermostat. Ironically, this control is usually taken more for granted than any other component in the entire system. Most homeowners just set their thermostat at a particular setting and never think twice about the many complex functions it must coordinate.

Thermostats can be either standard or programmable, Figures 2-1 and 2-2. Both the standard and programmable thermostats are low voltage (24 V); however, line voltage (120 V) thermostats are also used in residential applications, mostly for electric heat. For this reason, all the basic principles discussed here also apply to line voltage thermostats.

Figure 2-1. Standard Thermostat. Courtesy, Honeywell Incorporated.

Figure 2-2. Programmable Thermostat. Courtesy, Carrier Corporation,, a Subsidiary of United Technologies Corporation.

THERMOSTAT FUNCTIONS

The thermostat performs many functions. Some of the most important functions are listed below.

- In order to maintain an even room temperature, the thermostat must control the furnace on and off times. The thermostat also needs to have a wide range of settings to meet individual comfort requirements.

- The thermostat must prevent over- or underheating by making sure the burners do not stay on or off too long. By accurately monitoring the on and off times of the furnace, the thermostat keeps the spread between the coldest and warmest room temperature readings within 1 °F. The range of room temperature is called temperature swing. For ideal comfort conditions, it is desirable to maintain a room temperature swing within 1 or 2 °F. The amount of swing depends upon the design conditions and the equipment selected.

- Since the heat loss from a particular room accelerates when extreme outside temperatures occur, the thermostat must be able to compensate for these fluctuating outside temperatures and maintain the desired room temperature. For instance, when the temperature outside is 40 °F, the furnace has longer off times. This is because the room heat loss is not as great at this temperature, and the amount of heat delivered to the room remains there for a longer period of time. However, when the outside temperature drops to -10 °F, the internal losses accelerate, and the furnace might have to run all the time in order to maintain uniform temperature conditions. Thus, the thermostat has to sense and react to many variables, including the inside room temperature, the outside temperature, and the amount of time the furnace actually runs.

- A thermostat must be able to compensate for vibration and thermal inertia.

- Other than setting the condition desired within the space, the user should not have to react to any of the variables listed above. It is the job of the thermostat to automatically respond to the variables.

THERMOSTAT COMPONENTS

A standard, or dial, thermostat is composed of four basic components: the thermometer senses the indoor room temperature; the dial allows the owner to select the desired temperature; a series of switches respond to selections of heating, cooling and fan control. The thermostat also contains internal components to compensate for other variables. The following paragraphs describe how these internal components work together.

BIMETALLIC ELEMENTS

A basic law in physics dictates that metals, when heated or cooled, react by expanding or contracting to some extent. This expanding and contracting action can be used to activate components. This is the basic principle of the bimetallic device.

The extent to which a metal expands or contracts depends upon the type of material used. One metal, when heated, might expand considerably while another metal, subjected to the same temperature and heating interval, might expand to a lesser degree. For example, under a specified temperature, brass contracts or expands a great deal, while the metal invar reacts only slightly.

When two dissimilar metals are bound together, they are called a bimetallic element, Figure 2-3. When two strips of two different metals are bound together, they are called a bimetallic strip. For example, when heating a strip made of brass and invar, Figure 2-4, the brass expands considerably more than the invar. The strip distorts, forming an arc with the brass on the outside, or larger, radius and the invar on the inside, or shorter, radius. When cooled, the brass shrinks more than the invar, becoming smaller and forming the short radius and curving the arc in the opposite direction. Anchoring one end of the bimetal strip, Figure 2-5, causes the free end to move up or down, depending on whether the strip is heated or cooled. Because of this action, a bimetal strip can be used as either a thermometer or a switch.

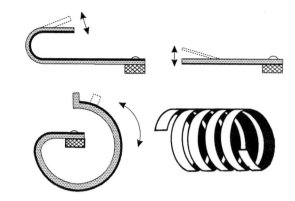

Figure 2-3. Various Shapes of Bimetallic Elements

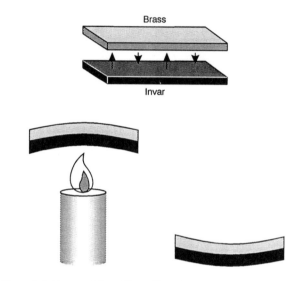

Figure 2-4. Brass and Invar Bimetallic Strip

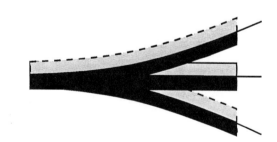

Figure 2-5. Anchored Bimetallic Strip

To function as a thermometer, a scale, calibrated in degrees of temperature, Figure 2-6, is positioned at the free end. The bimetal element directly reads the temperature of the surrounding air. Or, the element can function as a switch by using the calibrated scale with a set of electrical contacts at the movable end of the bimetal. The bimetal makes or breaks these contacts as it moves, Figure 2-7. Consequently, when the temperature rises or falls, the bimetal can open or close an electric circuit to a predetermined set point.

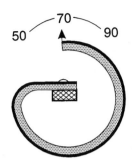

Figure 2-6. Calibrated Bimetallic Strip

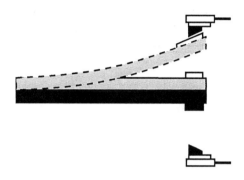

Figure 2-7. Bimetallic Strip Making/Breaking Electrical Contact

MERCURY SWITCHES

Besides functioning as a thermometer, the coiled bimetal can also be used to open and close another type of switch, known as a mercury switch, at a certain preset temperature. The mercury switch controls the on and off cycles of a furnace.

A mercury switch is a small glass enclosure with two wires or electrodes inserted in one end, Figure 2-8. Inside the bulb there is a bubble of mercury. Mercury is a liquid metal that flows back and forth in the bulb due to gravity. Mercury is also an excellent conductor of electricity.

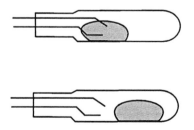

Figure 2-8. Mercury Switch in On and Off Positions

Understanding a mercury switch is fairly simple. For example, if the bulb is tipped away from the electrodes, the mercury flows to the opposite end, breaking contact.

As a result, the switch opens. On the other hand, if the mercury switch tips toward the electrodes, the mercury flows to that end and encloses both electrodes. This allows electrical current to flow in the same manner as a manual switch or any other type of switch.

By locating the bulb on top of the coiled bimetal, so the mercury flows back and forth as the coil rotates, the furnace circuit can energize or de-energize as the bimetal responds to changes in room temperature. When the temperature decreases, the coil contracts and rotates to the left, tipping the mercury bulb towards the electrodes and turning the furnace on, Figure 2-9. Conversely, when the furnace supplies warm air to the living space and the temperature increases, the coil unwinds in the opposite direction, tipping to the right. The mercury flows away from the electrodes, breaking the circuit and stopping the furnace.

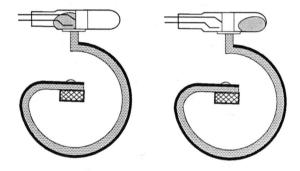

Figure 2-9. Mercury Switch Attached to Coiled Bimetal

MAGNETIC SWITCHES

Magnetic force is another way of energizing and de-energizing a furnace circuit in order to make and break contacts. In this case, the bimetal has an electrical contact on the end that moves when the temperature changes. Also located in the thermostat is a second fixed contact which includes a permanent magnet.

When the room temperature falls below the comfort setting, the moving contact approaches the fixed contact. The attraction of the magnet snaps the contacts together, completing the circuit and starting the furnace. In order to break contact, the bimetal must move in the opposite direction with enough strength to overcome the magnetic force holding the contacts together. In order for dependable operation to continue, this type of thermostat must be kept dirt- and dust-free, as these substances can foul the contacts and weaken the magnetic field, Figure 2-10.

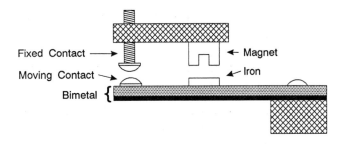

Figure 2-10. *Magnetic Switch*

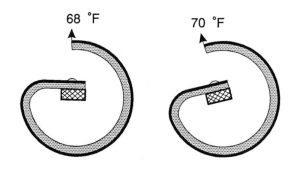

Figure 2-11. *Rotating Centerpoint of Bimetal*

SETTING THE COMFORT LEVEL

Every individual has a different opinion as to what temperature is most comfortable. Consequently, it is impossible for one room temperature to be perfectly comfortable to the variety of people occupying that space. With this in mind, the heating/cooling service technician's goal is to set the system to maintain a temperature that is reasonably consistent and comfortable to most people.

Usually the comfort level of a building or office is determined according to a combination of economic requirements and average comfort settings. In a home, however, the residents can set the design temperature to their personal comfort level. Therefore, a dial adjustment, or a programmable control, is often provided in the residential thermostat to allow individual selection of the most satisfactory room temperature. In some installations, the temperature is set in each individual room by means of damper controls and piping.

The thermostat is set by changing the temperature point at which the thermostat makes or breaks contact. The dial adjustment, or programmable control, is attached to the center, or anchor, point of the bimetal, Figure 2-11. When the desired temperature is changed, for example decreased from 70 to 68 °F, the centerpoint rotates slightly so the bimetal tilts a little farther away from the heating, or make, point of the bulb (the opposite reaction would occur if the thermostat setting was turned up). Burner cycles and running time remain the same; only the temperature maintained in the space changes. With this very simple adjustment, an individual can select and maintain the desired temperature within very narrow limits.

With continuous blower operation, the thermostat normally holds the room temperature within a range of plus or minus 1-1/2 °F, with 8 to 10 cycles/hour. Using short, but frequent, on times maintains the space temperature within very narrow limits, even in mild weather. Increasing the frequency of the on times makes it easier to maintain a minimal temperature swing above or below the set point of the thermostat.

With continuous blower operation, the thermostat normally holds the room temperature within a range of plus or minus 1-1/2 °F, with 8 to 10 cycles/hour. Using short, but frequent, on times maintains the space temperature within very narrow limits, even in mild weather. Increasing the frequency of the on times makes it easier to maintain a minimal temperature swing above or below the set point of the thermostat.

DIFFERENTIAL

The term differential refers to the temperature difference that occurs between the time the furnace turns on and off. Maintaining a narrow differential on a thermostat is essential in providing a consistently comfortable environment.

Under ideal conditions, a good thermostat maintains the room temperature within a narrow range. However, the lightweight bimetallic coil is so sensitive that vibrations can upset it enough to cause the mercury bulb to move back and forth, alternately making and breaking the contact. On a furnace, this can cause erratic cycling, which produces chatter.

Vibration cycling can be conquered by creating a slight hump in the middle of the mercury bulb, Figure 2-12. This helps curb the normal gravity flow of the mercury. In order to make or break the circuit, the mercury must first pass over this hump, which slows down the mercury's movement. This creates a temperature differential that causes a furnace to turn off at a higher temperature point than the point at which it turned on.

Figure 2-12. *Humped Differential Bulb*

Another factor is the weight of the mercury itself. The additional weight on the bimetallic coil requires the bimetal to twist further to make or break the circuit than is required by temperature alone. Thus, the temperature has a 2 °F differential due to the combination of the weight of the mercury and the hump in the bulb. This lag compensates for vibration in a room and prevents the mercury from erratically making or breaking the circuit.

HEAT ANTICIPATORS

A heat anticipator is a small wire resistor built into the thermostat. This component turns the burners off before a room actually reaches its temperature set point. This is necessary, because when a furnace is turned on, it first needs to heat up the metal in the heat exchanger and ductwork, as well as the air passing over the heat exchanger. As a result, the furnace does not reach its heating capacity immediately. This delay in heating up is called thermal inertia and can further increase the room's temperature differential. Including a heat anticipator in the installation helps compensate for thermal inertia.

Once the thermostat senses the cut-off temperature, it turns the burner off. However, the elements in the heating system retain a certain amount of heat. This heat is delivered to the room for a short period of time, causing the room to overheat even though the burners are off. Consequently, when the thermostat signals a unit to come on, there is a time lag before the unit begins delivering air at the proper temperature. This allows the temperature of the heated space to fall below the desired comfort setting.

The heat anticipator prevents overheating of the space by turning off the unit slightly ahead of time. Thus, it 'fools' the thermostat into believing the correct room temperature has been reached. Otherwise, surplus heat would be delivered to the room or space after the burners shut off. The remaining heat in the exchanger and duct system brings the space temperature only slightly above the set point of the thermostat.

The heat anticipator is located near the bimetal element and is wired in series with the gas valve or oil burner relay. When the anticipator is energized, these other components energize at the same time. Since it is a resistor, the heat anticipator gives off heat which is added to the room temperature sensed by the thermostat's bimetal element. This additional heat causes the bimetallic coil to rotate more than usual in response to the actual room temperature. As a result, the burners turn off before the room actually reaches the thermostat set point.

Many people sense a 2 °F differential and find it uncomfortable. A heat anticipator compensates for this differential by overriding the thermostat, in case the thermostat is overcome by vibration, and offsetting thermal inertia. This brings the thermostat back to a point where it can control room temperature, to within less than 1 °F, solving both the vibration and thermal inertia problems.

Some heat anticipators are fixed resistances, which means they are matched to the specific equipment to which the thermostat is applied. This type of heat anticipator cannot be adjusted. Most thermostats have an adjustable anticipator, so the resistance and heat produced meet the varying needs of the application. In the low voltage control circuit, the resistance is matched to the amount of current passing through the gas valve, oil burner relay, or other components so that the exact amount of heat is delivered by the anticipator. The more heat supplied to the bimetal, the shorter the cycle. Conversely, less heat produces a longer cycle.

Simply described, a heat anticipator is a piece of coiled wire with a movable lever connected to the transformer at one end. This lever selects the amount of resistance necessary. The lever is connected electrically to the gas valve control circuit at the furnace. The length of wire determines the total amount of heat produced and the amount of electricity or current passing through it. The greater the length of wire (resistance), the more heat produced by the anticipator. For example, if the control circuit draws 0.4 A, and the selector lever is moved to 0.25 A, the anticipator produces more heat with more resistance and shortens the burner cycle. Figures 2-13 and 2-14.

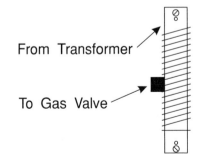

Figure 2-13. *Heat Anticipator as a Resistor*

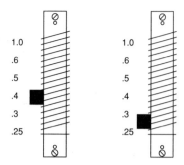

Figure 2-14. Setting the Anticipator

Adjustable heat anticipators are useful for several reasons. One reason is that some people prefer their burner to cycle on and off more frequently. Raising the setting lengthens the cycle, and lowering the setting shortens the cycle. Another reason is that gas valve manufacturers often make valves with different amperages, with a higher amperage valve requiring fewer turns of a resistor to generate the heat required to produce a 3-minute cycle. Another advantage of an adjustable heat anticipator is that even though a gas valve draws its normal 0.20 A, if more control relays are added to the same circuit, the amperage draw of the total circuit changes. An adjustable heat anticipator compensates for this and still generates enough heat to produce a 3-minute cycle.

The majority of gas valves are marked for a 0.20 A rating and anticipator setting. Almost all furnaces have a plate or sticker displaying manufacturer's information concerning the heat anticipator's proper setting. If this information is not available, the general rule is to set a gas furnace at 0.25 A, an oil furnace at 0.45 A, and an electric furnace between 0.20 A and 0.40 A. In some cases, it is necessary for a service technician to set and reset the heat anticipator until the desired setting is obtained, see Figures 2-15 and 2-16.

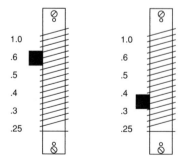

Figure 2-15. Different Gas Valves

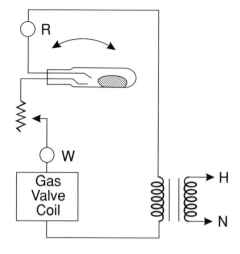

Figure 2-16. Heating Thermostat Circuit with Heat Anticipator

COOLING FUNCTION

Just as heat anticipators adjust a thermostat to heat a particular living space, there are also cooling functions contained within specific thermostats that cool a living space.

In the cooling mode, the bimetallic coil is positioned to respond to an increase in heat so the living area can be cooled. The air conditioning, or cooling cycle, activates when contact is made at a certain point. When the area cools sufficiently, the bimetallic coil reverses this motion, causing the unit to cycle off.

The mercury bulb for cooling may be designed exactly as it is for heating—with a hump in the middle to provide a 2 °F differential. The cooling function also comes equipped with an anticipator, since the principle of thermal inertia remains the same. The time lag in the cooling mode, however, is opposite that of the heating mode. The lag allows the evaporator and air ducts to cool down so they deliver cool air to the space.

To maintain an even space temperature, the cooling system must come on slightly before the space temperature reaches the temperature set point. The swing is plus or minus 2-1/2 °F, with 3 to 4 cycles/hour. The cooling anticipator differs from the heating anticipator in that it is a fixed resistor matched to the compressor contactor and therefore it cannot be adjusted or changed in the field.

The cooling resistor is wired in parallel with its mercury bulb, Figure 2-17. The resistance through the mercury switch is practically zero when it is closed and the compressor is running. As current takes the path of least resistance, it flows through the switch, rather than through the anticipator. With no current flowing, the anticipator does not produce any heat. All of the current flows through the compressor contactor, pulling in the contactor and operating the compressor.

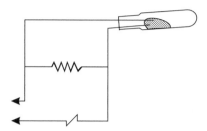

Figure 2-17. Cooling Resistor Wired in Parallel with Mercury Bulb

The resistor is in series with the compressor contactor when the cooling demand is satisfied and the mercury switch is open. This current is sufficient to produce heat in the resistor; however, due to the large voltage drop across the resistor, there is not enough voltage remaining to pull in the compressor contactor, even though some current is flowing through it. The reason for this is that a compressor has a much greater voltage demand in cooling than in heating.

An example of this current passing through a contactor, but not in sufficient quantities to be pulled in, is similar to a 100-W bulb that does not light because of the very small resistance, while a 25-W bulb lights because of a much greater resistance. Comparing a 100-W bulb to a compressor contactor and the 25-W bulb to a cooling anticipator, when the switch is energized (calling for cooling), the 100-W bulb glows brightly when it receives full voltage.

During the off cycle, the cooling anticipator generates heat and 'fools' the bimetal into thinking the temperature in the space is higher than it actually is. This causes the mercury bulb to tilt, the circuit to close, and the compressor to start before the temperature actually reaches the set point. Thus, the cooling anticipator compensates for the differential in the mercury switch and anticipates the need for cooling. Anticipators are extremely valuable for both the heating and cooling modes. They allow the thermostat to maintain the temperature of a living space within very narrow limits, usually plus or minus 1 °F.

One last note about the cooling anticipator. The on/off time of the air conditioning unit is shortened, because the cooling anticipator starts the air conditioning system slightly ahead of time. This helps to control the humidity of the space. Humidity is defined as the moisture level of a space and is an important factor in maintaining a cooling comfort level. The importance of a proper humidity level is discussed in later chapters.

COMBINATION HEATING/COOLING THERMOSTATS

Thermostats can be built with two bulbs, one for heating and one for cooling. They can be combined into a single thermostat by mounting one bulb on top of the other. As this configuration adds considerable weight to the system, for which the bimetal coil must compensate, this is not a practical solution. The best arrangement involves placing a cooling electrode at one end of the bulb and a heating electrode at the other end of the same bulb, then adding a third, or common, electrode.

In the combination thermostat, when the bulb is tipped in one direction, the mercury completes the circuit through the cooling and common electrodes. When it is tipped in the opposite direction, it completes the circuit through the heating and common electrodes. In this arrangement, it is not possible for the thermostat to call for heating and cooling at the same time.

One possible location for the heating and cooling anticipators is immediately beside the bimetallic coil. This is so the coil has the proper anticipator for the function selected. The basic wiring for each anticipator is the same as that shown in the individual wiring diagrams, but instead, both are combined in one circuit.

Having both heating and cooling in the same circuit presents a potential problem. When the heating bulb is energized, the furnace comes on and the heating system then operates. When the heating load is satisfied, the bulb switches back to shut off the heating system. In doing so, it turns on the cooling system and subsequently reduces the heat in the space. The two bulbs are constantly switching back and forth from heating to cooling. This is not acceptable for proper temperature control and operation.

The solution to this problem is a heat-off-cool switch. This switch places the system either in heating or cooling, without switching back and forth. Figure 2-18 shows a typical heating-cooling internal thermostat circuit with both heating and cooling anticipation. This circuit also shows a heat-off-cool switch. For heating, the furnace blower cycles on the furnace fan control with the burners. For cooling, the blower cycles through a cooling relay with the compressor.

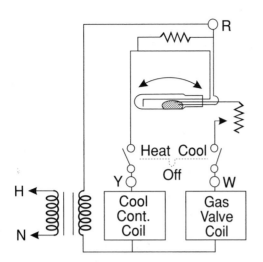

Figure 2-18. Heating-Cooling Anticipation Thermostat Circuit with an Off Switch

The cooling circuit cannot be energized when the switch is in the heating position, the heating load is satisfied, and the bimetal swings over to cooling. This is due to the open switch in the thermostat. The same rule applies when the switch is in the cooling position and the cooling load is satisfied; the heating circuit cannot be energized. This is considered to be a double-pole, double-throw (DPDT) switch with an off position in the middle. The off switch prevents the heating and cooling circuits from being energized.

THERMOSTAT SUBBASES AND FACES

Thermostats come in many different shapes and forms, but their basic function never really changes. The portion of the thermostat that is leveled and mounted on the wall is referred to as the subbase. The subbase contains all the wiring connections to the thermostat. Connecting the face or cover to the subbase completes any electrical contacts the face requires.

The subbase contains the blower switch as well as the basic heat-off-cool switch. The purpose of a subbase is to allow the service technician access to all of the wiring without disturbing the basic thermostat. Many subbases contain printed circuits to complete the internal wiring of the thermostat. In this case, it is only necessary to connect the four or five thermostat leads to the designated terminals. The thermostat always includes details concerning the installation and wiring, and the service technician should have access to them.

FAN CONTROL CIRCUITS

Many thermostats also have a switch for the selection of blower operation. This is sometimes confusing, as the selection may be marked on and off; on and auto; cont. and auto; or cont. and int. The following is a breakdown of what these markings represent:

■ On or cont. means that the blower runs continuously, independent of the burner cycles.

■ Auto, off or int. positions indicate that the blower cycles on and off with the burners.

In the auto or off position, the blower does not come on until the burners have been on for several seconds. This allows the heat exchanger and ductwork to warm up. Then the blower remains running for several seconds after the burners are off. This allows some of the heat to dissipate in the heat exchanger, preventing it from getting too hot.

Without a blower switch, the thermostat can only cycle with the burners on, through the fan control. If continuous blower operation is desired, additional wiring is necessary at the furnace. Figure 2-19 shows a typical heating-cooling internal thermostat circuit with both heating and cooling anticipation. This circuit has a heat-off-cool switch plus a fan selection switch for auto-on. The switch is shown in the auto position, where the fan cycles on both heating and cooling. The fan runs constantly when in the cont. position.

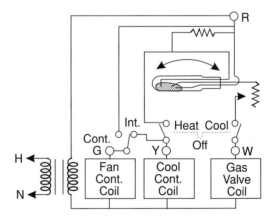

Figure 2-19. Heating-Cooling Circuit with Fan Control

Most thermostat manufacturers label their terminals with the same letter coding that corresponds to the wire color used. Several widely used thermostats are listed with their coding:

Manufacturer	Type	Common	Cooling	Heating	Fan
General	T91	V	C	H	F & G
General	T199	R	Y	W	G
Honeywell	T834	R	Y	W	G
Honeywell	T87	R	Y	W	G
Cam-Stat	T17	R	Y	W	G
Cont. Corp.	360	R	Y	W	G
White-Rodgers	IF56	RC & 4	Y	W	G

Some thermostats have additional terminals for two-stage heat (Wl-W2), two transformers (Rh and R) and other options. Figure 2-20 shows the internal wiring of a heating-cooling thermostat with fan control selection and the provision for using two transformers. When only one transformer is used, the R and Rh terminals are jumpered, and the thermostat functions similar to the one shown in Figure 2-19. The jumper is removed when two transformers are used. In this case, the internal switching (by selecting heat or cool) uses one transformer for heating and the other for cooling.

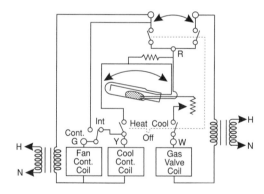

Figure 2-20. *Heating-Cooling Fan Control Circuit Using Two Transformers*

INSTALLING CONVENTIONAL THERMOSTATS

The application and installation of a thermostat directly affects its performance and efficiency. A poor comfort control complaint may simply be due to incorrect mounting of the thermostat.

CORRECT THERMOSTAT LOCATIONS AND APPLICATIONS

The thermostat needs to meet certain criteria, or else customer complaints may result. These criteria are as follows:

■ The thermostat must be leveled accurately, because the mercury switch depends on gravity to function properly, Figure 2-21. If the thermostat is tilted, the dial adjustment setting becomes inaccurate and, as a result, the thermostat does not function correctly.

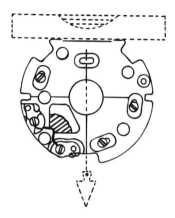

Figure 2-21. *Leveling a Thermostat*

■ The thermostat should be positioned about 55 inches from the floor. Since cool air travels downward and warm air travels upward, the thermostat should be located where it can give an average reading of the space concerned.

■ The thermostat should be located in the space most frequently used (usually the living room), preferably on an inside wall where it can sense an average temperature of the conditioned space.

INCORRECT THERMOSTAT LOCATIONS AND APPLICATIONS

There are many locations and applications of a thermostat that should be avoided, as all them result in poor control of the average temperature within the room. The thermostat should not be mounted in the following areas:

■ In the direct path of a window near direct sunlight or an open space, as this can result in a false reading.

■ In any area that might be prone to any type of vibration (i.e., near a door that is constantly being opened or closed). Continued vibration may cause the thermostat to be thrown out of calibration.

■ In or around any confined space where the thermostat cannot receive free air flow of the conditioned space. This type of location gives the thermostat a reading of that area only, not of the total space.

■ Near any type of chimney or ductwork, because the heat produced by these may cause a false reading. By the same token, the thermostat wiring should always be enclosed, not sensing drafts of any kind, especially through the hole into which it is installed.

■ On an outside wall, because the thermostat is affected by the heat loss or gain through the wall.

All of these conditions must be avoided in order to control and receive proper readings and signals from the thermostat, and to provide proper heating or air conditioning comfort.

PROGRAMMABLE THERMOSTATS

Programmable thermostats are now replacing many of the older dial thermostats. They are extremely popular, because the owners can program the thermostat to fit their specific lifestyles. For example, if the owners wake up at 7:00 a.m., they can set the thermostat to turn on the heat at 6:45 a.m., so the living space will be warm by the time they get out of bed. These thermostats make maintaining a comfortable environment practically trouble-free.

The programmable thermostat used in the following examples is typical of most programmable thermostats. However, every manufacturer has a different method of programming a particular thermostat. It is very important for the service technician to read all manufacturer literature concerning the programmable thermostat that is to be installed, as this literature contains valuable information concerning operation of the unit.

INSTALLATION

When installing a programmable thermostat, it is always necessary to read and follow all the manufacturer instructions carefully. Also, to prevent electrical shock or equipment damage, it is important for the service technician to always disconnect the power supply at the furnace or boiler before installing the thermostat. A system using a two-stage transformer may require turning off two switches or disconnects. The subbase should then be removed from the wall, and at this point, it is important for the service technician to write down the letter or number on the wiring terminal and its corresponding wire color as the wire is removed. This helps avoid confusion when the subbase is placed back on the wall.

It is important for the service technician to follow all local electrical codes when installing a thermostat. Also, the wiring should be restricted to a recessed area surrounding the terminals, to ensure thermostat-subbase contact.

CHECKOUT

Once the programmable thermostat is installed, the service technician should perform a checkout procedure. This procedure ensures the unit is functioning correctly.

When checking the thermostat heating function, perform the following steps:

1. Move the fan switch to the auto position.
2. Move heat-cool switch to the heat position.
3. Set the thermostat to a temperature approximately 10 °F above the living space temperature. At this point, the heating system should start and the fan should run.
4. Reduce the thermostat setting to about 10 °F below the living space temperature. The heating should end and the fan should turn off.

When checking the thermostat cooling function, perform the following steps.

CAUTION: Never operate the cooling system unless the outdoor temperature is above 50 °F.

1. Move the heat-cool switch to the cool position.
2. Move the fan switch to the auto position.
3. Set the thermostat to a temperature approximately 10 °F below the living space temperature. The cooling system should start and the fan should run.
4. Increase the thermostat setting to approximately 10 °F above the living space temperature. The cooling system and the fan should both stop.

When checking the thermostat fan function, perform the following steps:

1. Move the heat-cool switch to off.
2. Move the fan switch to the on position. The fan should run continuously. When the fan is moved to the auto position, the fan cycles with either the cooling or heating system.

PROGRAMMING

In order to program the thermostat, 24-V power must be restored. When the power is restored, the thermostat should display the room temperature and a certain time (i.e., 12:00). Sometimes, this temperature and time will flash on and off until the unit is programmed. When programming the thermostat, the service technician should never use a pen or pencil, as these can damage the keyboard. A fingertip or other blunt tool should always be used.

The actual programming of the thermostat differs from one manufacturer to another; therefore, it is necessary for the service technician to always read the manufacturer's literature before programming. The service technician may set the current day and time, but the owners must decide on the actual thermostat settings, based on their own comfort levels.

REVIEW QUESTIONS

1. What is a thermostat?
2. What are the main components of a thermostat?
3. How is a bimetal used?
4. What is the advantage of using mercury in a thermostat?
5. What is the purpose of the heat anticipator?
6. What is a mercury switch?
7. What is a magnetic switch? SNAP ACTION
8. What is meant by differential? TEMP. DIFFERENCE BETWEEN ON AND OFF
9. Explain why a heat anticipator must be built into a thermostat. TO PREVENT OVERHEATING THE SPACE
10. What is a cooling anticipator? A FIXED RESISTER IN PARALLEL WITH CSTAT
11. Is there current through the compressor contactor, when energizing the cooling anticipator? YES BUT NOT ENOUGH TO OPERATE THE CONTACTS
12. Why is cooling anticipation important? MAINTAIN AN EVEN SPACE TEMPERATURE
13. What is the purpose of the thermostat heat-off-cool switch? PREVENT AUTOMATIC SWITCHING BETWEEN HEAT +COOL IN A COMBO THERMOSTAT
14. What cycles the furnace blower on heating? HSTAT
15. What cycles the furnace blower on cooling? CSTAT TO WIRES
16. Describe the function of a subbase. ALLOWS ACCESS WITHOUT DISTURBING THE SETTINGS
17. What is a programmable thermostat?
18. What are the advantages of a programmable thermostat?

Chapter 3

Heating Fundamentals

As furnace designs are based upon certain heating principles, such as the nature of heat, heat measurement, and combustion, it is necessary to understand these principles before studying the specific components of a heating unit.

PRINCIPLES OF HEAT

As stated in Chapter 1, heat is a form of energy that cannot be created or destroyed. However, one form of energy can be transformed into other forms of energy, and energy can be transposed from one place to another. For instance, fuel, as a source of energy, can be burned and transformed into other forms of energy: heat and light.

Cold, the opposite of heat, is not really a separate entity but is a relative term. Cold is actually just a lesser heat content, or an absence of heat. In fact, substances contain heat all the way down to a temperature of -460 °F, called absolute zero.

MATTER

Heat is also associated with matter. Matter, as stated previously, consists of small particles called molecules that are constantly in motion. This motion is called kinetic energy. The speed at which the molecules move determines the state—solid, liquid, or gas—of the substance. In a solid, the molecules merely vibrate; in a liquid, they move freely; and in a gas, they move rapidly. The addition of heat energy increases the motion of the molecules and thus their kinetic energy, or motion, can be sensed by touch; this is called temperature. Temperature is the measure of the intensity of heat.

The points at which water (matter) changes from a solid to a liquid and back again are the basic reference points for temperature measurement. These reference points are taken at sea level, because at this level, the atmospheric pressure is constant. This constant is 14.7 psia

(lbs/ square inch absolute). At a higher level, such as on a mountain, it is impossible to boil water to a temperature of 212 °F (the boiling temperature of water), because the air is thinner on a mountain top than it is at sea level. Because the atmospheric pressure varies from one high altitude to another, a constant is needed; hence, sea level is the constant.

On the Fahrenheit scale, water freezes at 32 °F and boils at 212 °F. These are fixed reference points. Between these reference points are 180 equal divisions, or degrees, which measure temperature change. On the Centigrade scale, the reference points are 0° for freezing and 100 °F for boiling. These reference points have 100 equal divisions, or degrees, in between, Figure 3-1.

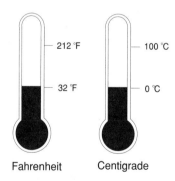

Figure 3-1. *Fahrenheit and Centigrade Thermometers*

HEAT ENERGY

British thermal units (Btu) measure the quantity of heat, called heat energy, in a substance. A Btu is the amount of heat required to raise the temperature of 1 lb of water 1 °F. A Btu measures the total heat content of a substance by volume. For example, suppose there are two pans of water; one is filled with 1 lb of water, and the other pan is filled with 10 lbs of water. Both pans have a temperature of 70 °F. Even though the temperature of both pans is the same, the 10-lb pan of water has ten times more heat energy than the 1-lb pan. Looking at it in terms of Btu or heat energy, it takes only 10 Btu to raise the 1-lb container of water 10 °F, but it takes 100 Btu to raise 10 lbs of water 10 °F. The formulas for these instances are:

10 °F x 1 lb x 1 Btu/lb = 10 Btu required

10 °F x 10 lbs x 1 Btu/lb = 100 Btu required

Keeping this simple formula in mind when discussing Btu is helpful for continued reference.

COMBUSTION

The primary purpose of a furnace is to transfer the heat caused by combustion to a living space. Consequently, a service technician must have a basic understanding of the process of combustion and how it takes place in a furnace. Combustion is the process of burning a fuel (such as gas or oil in a furnace) in order to produce two other forms of energy: heat and light. Combustion is called a high speed oxidation process. All this means is that fuel and oxygen produce heat when they are combined and burned; this is combustion. The heat produced by the reaction continuously relights the unburned portion of the fuel, resulting in a constant burn.

Combustion results in a flame. The flame occurs when a combustible (i.e., oil or gas) combines with atmospheric oxygen and is heated to its ignition point (the point at which it ignites and burns). The reaction produces heat and light in the form of a flame. Flames are discussed in more detail later in this chapter.

The combustion triangle, Figure 3-2, illustrates the three items mentioned that must all be present in order to produce a flame. These three items are as follows:

Fuel. Any substance that burns can be used. Natural gas, LP gas or oil are the usual fuels found in residential heating. Fuels are discussed in more detail later in this chapter.

Heat. Each substance has a different ignition point. To raise the temperature of the fuel to this point, enough heat must be supplied. In most cases, a pilot light (constantly burning) produces the heat for ignition.

Oxygen. A sufficient amount of oxygen must be present in order to support combustion. And, as oxygen is consumed in the combustion process, it must continually be replaced. Surrounding air and forced air (air drawn in by a blower) are the two sources used to supply oxygen.

The need to have oxygen present for combustion is demonstrated by trying to burn fuel oil in an open pan. Oil

is a combustible and as such, a match or other flame can easily bring it to the ignition point. However, because oil is heavy, it must be atomized (broken into tiny droplets and mixed with air). If oil is not atomized, it does not burn. Even if a very high flame causes it to ignite, it immediately burns itself out if there is a lack of oxygen.

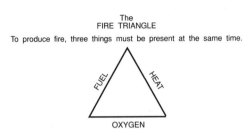

The
FIRE TRIANGLE
To produce fire, three things must be present at the same time.

FUEL   HEAT

OXYGEN

If any one of the three is missing, a fire cannot be started or, with the removal of any one, the fire will be extinguished.

Figure 3-2. *Combustion Triangle*

FUELS

A number of common fuels are used in residential heating, each having different characteristics and requirements. The service technician must understand each fuel type in order to correctly adjust the furnace burners.

NATURAL GAS

Natural gas is the most widely used fuel. It is made up of about 99 percent methane. Methane, in turn, is composed of carbon and hydrogen, and is called a hydrocarbon, which is an excellent fuel. The heating value of natural gas ranges from 950 to 1,125 Btu/cubic foot (ft^3). In most circumstances, however, it is considered to be about 1,000 Btu/ft^3. Therefore, a furnace rated at 110,000 Btuh (Btu/hour) burns 110 ft^3 of gas/hour. Natural gas is a relatively slow-burning gas with a maximum theoretical flame temperature of 3,550 °F. It has a specific gravity of 0.65—literally lighter than air, which has a specific gravity of 1.00.

MANUFACTURED GAS

Another less commonly used gas is manufactured gas, which is a by-product of other substances. It is composed of hydrogen, methane, carbon monoxide and small quantities of carbon dioxide, oxygen and nitrogen. Its heating value is about half that of natural gas, approximately 500 to 550 Btu/ft^3. As manufactured gas is a relatively fast-burning gas, the speed of burning increases as the hydrogen content increases. The maximum theoretical flame temperature is about 3,560 °F, and its specific gravity averages 0.45.

Some areas pipe in natural gas and mix it with locally-produced manufactured gas. This mixture averages from 700 to 900 Btu/ft³, and it has a maximum flame temperature and specific gravity very close to that of natural gas. In this case, an orifice of a different size is usually the only change necessary to achieve satisfactory combustion.

LPG

Areas or sites not served by natural gas pipelines (i.e., rural areas and mobile homes), often use liquefied petroleum gas, or LPG. Butane and propane are two common LP gases, and they are comprised of straight hydrocarbons compressed into a liquid state. They create their own vapor or gas pressure when their temperature is above the boiling point (butane +31 °F, propane -43 °F) and produce about 21,600 Btu/lb in the vapor state. Vaporized butane, which is twice as heavy as air, has a specific gravity of 2.01 and contains 3,200 Btu/ft³. Vaporized propane has a specific gravity of 1.52 and contains 2,500 Btu/ft³. The flame temperature of both gases is about 3,650 °F.

LP gases are sometimes mixed with manufactured gases or air. These mixtures can result in a specific gravity of less than air, although they may be diluted to the point where they contain only 550 to 550 Btu/ft³. This is important, because any gas heavier than air requires special safety controls, including 100 percent shut-off.

OIL

Oil-fired furnaces use several different grades of oil, which are classified according to their viscosity. Viscosity describes an oil's resistance to flow—in other words, its thickness. The thicker an oil, the more resistant it is to flow and the higher its viscosity. For instance, molasses has a higher viscosity than ordinary vegetable oil.

Oils are graded according to their viscosities on a scale from 1 to 6, excluding 3. Oils No. 1 and No. 2 are thin oils, like diesel. Oils No. 4 is a little thicker, No. 5 oil is very thick, and No. 6 oil, when cold, resembles tar. The higher the viscosity, the more Btu/gallon the oil contains. For example, the most commonly used oil, No.2, has a Btu content of 144,000, whereas No.6 oil contains 152,000 Btu.

When using fuel oils, a service technician must be able to distinguish between two important temperatures: the flash point and the fire point. The flash point is the temperature at which the volatile elements in the fuel flash when exposed to an open flame. The fire point is the temperature at which the oil burns continuously.

FUEL IN COMBUSTION

There is a significant difference between the burning properties of various fuels. The type of fuel used in a furnace also determines the quantity of air required for combustion. Air in excess of the minimum required quantity is necessary for complete combustion. The volume of excess air required depends on factors such as fuel properties, size and design of the combustion chamber, maximum furnace temperature, and thoroughness of the air/volatile gas mixing.

An inherent problem with supplying excess air to ensure complete combustion is that operating efficiency is reduced to a certain degree. The excess air lowers the furnace temperature and absorbs heat that should be more properly utilized for generating steam.

Furnace temperature influences the fuel ignition and the combustion rate as well. In order for combustion to take place, the mixture of air and fuel combustibles must be raised to its ignition temperature. Once the firing point of ignition is reached, the furnace must be capable of rapidly supplying oxygen in the required amount to sustain complete combustion. When the furnace temperature exceeds the ignition requirement, temperature is relatively insignificant because it exerts little influence on the oxidation process of combustion.

PRODUCTS OF COMBUSTION

Hydrogen and carbon atoms are present, in varying amounts, in all fuels. These atoms are called hydrocarbons.

Hydrogen mixes readily with air and burns at a faster rate and at a lower temperature than carbon. It has a bluish color and burns first, using the air it needs for complete combustion. The carbon particles are not completely burned until they reach the outside edge of the flame, where they can obtain the required amount of air for complete combustion. The unburned carbon particles, which burn more slowly but at a higher temperature, produce the bright light normally associated with an open fire.

When an open fire is examined closely, a small blue flame at the base (caused by the hydrogen) is seen, surrounded by a much larger area of yellow (the unburned carbon). Since the hydrogen burns first in the inner core, it uses up all the available oxygen. As a result, there is no oxygen left in the flame to completely burn the carbon. The carbon does not burn completely until it reaches the outermost edge of the flame, where it obtains the oxygen it needs from the surrounding air.

During combustion, a chemical change occurs and most of the elements are rearranged. The fuel is not destroyed; it is a source of energy, and according to the rule, energy can be transformed but not destroyed. The fuel contains hydrogen (H) and carbon (C) atoms, while the air contains oxygen (O) and nitrogen (N) atoms. The products of complete combustion are carbon dioxide (carbon + oxygen, or CO_2) and water (hydrogen + oxygen, or H_2O). The nitrogen from the air passes through the combustion process unchanged. None of the products of complete combustion are harmful. Even though different amounts of air are required for complete combustion of gas or fuel oil, the resulting products of combustion are the same.

COMPLETE COMBUSTION

Complete combustion is the burning process of a certain quantity of fuel that is completely consumed in an oxygen-rich environment. It should be noted, however, that perfect combustion cannot be achieved under normal operating conditions. Perfect combustion can only be attained in a laboratory situation where air/fuel mixtures and ignition temperatures can be controlled with a high degree of precision.

Complete combustion of any fuel requires approximately 10 ft^3 of air for every 1,000 Btu of heat contained within the fuel. In the case of natural gas, there are approximately 1,000 Btu/ft^3; theoretically, complete combustion requires 10 ft^3 of air for 1 ft^3 of gas. In a gas furnace, 40 to 50 percent excess air is desirable. Therefore, good design allows for between 14 and 15 ft^3 of air/ft^3 of natural gas.

Burning LPG—for example, butane, which contains 3,260 Btu/ft^3—theoretically requires about 33 ft^3 of air. In contrast, 1 ft^3 of gas requires about 48 ft^3 of air in order to burn. This is the reason the LPG burners must be designed differently than for natural gas.

When using No. 2 fuel oil, the ratio is 1 lb of fuel oil plus 14.4 lbs of air (11.1 lbs nitrogen and 3.3 lbs oxygen) with a flame for ignition. Complete combustion yields 11.1 lbs of nitrogen, 3.2 lbs of carbon dioxide and 1.1 lbs of water. In terms of gallons, 1 gallon of fuel oil requires 1,540 ft^3 of air.

INCOMPLETE COMBUSTION

Incomplete combustion occurs when the fuel supply is not entirely consumed during combustion. This is caused by an oxygen-starved gas flame. Incomplete combustion creates soot, as well as a number of harmful elements, including carbon monoxide, aldehydes, ketones, and oxygen acids.

Incomplete combustion can also occur if a flame is cooled below its ignition point. This is demonstrated by holding a cool glass ashtray over a candle. Soot immediately forms on the bottom of the ashtray, indicating that part of the carbon did not burn. Soot is composed of semi-solid particles of unburned carbon and is always an indication of incomplete combustion. In sum, incomplete combustion occurs when a flame is cooled below its ignition point (called impingement of the flame) or from insufficient oxygen.

FLAMES

Yellow and blue are the two basic types of flames. There are variations of these flames, and these variations are due mostly to either primary or secondary air.

YELLOW FLAME

An open fire, such as a candle or a bonfire, is most readily associated with a yellow flame. Unburned carbon produces the yellow color. As stated previously, however, these particles do burn completely once they reach the outer edge of the flame and can obtain the additional oxygen required for complete combustion. A yellow flame requires a lot of space in which to burn. If confined, the flame is subject to impingement which in turn, results in incomplete combustion.

Since an oil burner does burn with a yellow flame, its characteristics should be understood. In an oil burner, the oil is atomized in the nozzle, which allows each drop to ignite. However, each droplet does not attain complete combustion until it reaches the outer edge, which results in its characteristic yellow flame.

BLUE FLAME

One significant type of blue flame in gas heating results from mixing about half of the combustion air required for the burning process with the gas before it is ignited. The flame produced by this method is blue rather than yellow, much smaller in size, has a higher temperature, gives out only a small amount of light, and is easily adjustable. After burning equal quantities of gas, the total amount of heat produced by a yellow and a blue flame is about the same.

The simplest demonstration of blue flame characteristics involves a Bunsen burner, found in most science labs. In a Bunsen burner, gas is introduced under pressure at the bottom of an upright tube. Just above the base, there is a collar containing slots and a cover that can be rotated to open all or part of the slots. Opening the slots allows a quantity of air, called primary air, to mix with the gas on its way up to the top of the tube, where the mixture is ignited. Even today, most burners in a gas furnace are based on this basic principle, because this system is efficient and easy to build.

When there is a proper amount of primary air, the flame has a bright blue inner cone surrounded by an outer mantle of light blue, Figure 3-3. This is the basic flame used in all gas furnaces. Combustion in this type of flame simultaneously occurs in two zones: on the surface of an inner cone and on the surface mantle. As such, all of the carbon particles remain in a gaseous state until they reach the point of ignition. The inner cone is composed of a gas-air mixture rather than unburned carbon. There is no yellow color, because there are no unburned carbon particles.

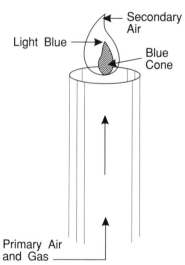

Figure 3-3. Primary Air Produces a Blue Flame

The primary air, mixed with the gas, supplies enough oxygen to support combustion on the inner cone while the secondary air (air that is added to the flame after ignition takes place) supplies the oxygen for complete combustion on the outer mantle. The flame remains blue, indicating no unburned carbon, as long as there is sufficient primary air. If the primary air is restricted, a yellow tip appears on the cone, due to the carbon particles not being completely burned until they reach the outer

edge of the flame. A yellow flame appears when primary air is completely shut off.

There are situations in which a blue flame changes to a yellow flame, indicating incomplete combustion. Causes of this are shown in Figure 3-4 and described below:

Lack of Secondary Air. Even though the flame is adjusted to receive a proper amount of primary air, if there is not enough secondary air, incomplete combustion results. In this case, the outer mantle does not have enough oxygen to burn all the gas.

Excess of Primary Air. Too much primary air distorts the flame and lifts it off the burner port. This causes a blowing noise and the flame could blow itself out.

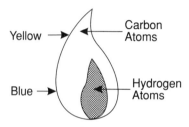

Figure 3-4. Blue Flame Changing to Yellow Flame

An air-fuel mixture using a greater air to fuel ratio results in a hotter fire, reduced fuel consumption, and harmful by-products from incomplete combustion not being discharged into the atmosphere.

HEAT TRANSFER
Another principle of heat energy used extensively in the heating and air conditioning industry is that heat always flows from a warmer body to a cooler body. This behavior allows heat to be transported from one place to another. Heat travels by three methods: conduction, radiation and convection.

CONDUCTION
Some materials conduct heat better than others. For example, if one end of an iron rod is placed in a fire, the other end quickly becomes hot, as the heat energy travels from the hotter end, through the rod, to the cooler end. This type of heat transfer is called conduction.

Other materials do not conduct heat as well. This is demonstrated by pouring boiling water into a cup. The cup gets warm and, if a metal spoon is placed in the cup, the opposite end of the spoon becomes hot. However, a wooden spoon placed in the same cup does not get hot on the far end, because wood resists the conduction of heat.

Conduction also occurs when different materials (including air) come in contact with each other. For example, the wall of a house can be composed of an air film on the inside wall, plasterboard or wood paneling, insulation, an interior air space, bricks or concrete blocks, and an outside air film. Heat travels, by way of conduction, through all of these materials. The direction of heat travel depends on the season and whether it is warmer inside or outside of the house.

RADIATION

Heat waves or rays transmit heat energy. This energy moves freely in space and has some unusual properties. When heat waves come in contact with a material, they may pass through the material as light. When this occurs, the material is not warmed to any degree. If the material has a smooth, bright surface, the rays may be reflected away from the material; if it has a dark or rough surface, the rays may be absorbed by the material. A material becomes hotter when it prevents the free passage of heat waves. In either case, some measure of radiant heat is always present.

Radiant heat always moves from the warmer to the colder body. For example, heat waves travel from a person's body toward a cold window, or from a fireplace to the walls of a room. While radiation is not considered a major factor in air conditioning, provision is made in commercial heat gain tables for the types of roof surfaces.

CONVECTION

Another heat principle that is useful in heating design is as air temperature increases, the volume of air expands. As the air expands, it gets lighter in weight, and subsequently, rises. Cooler, heavier air settles in the bottom of the space. The warmer air gets, the more rapidly it moves, and when it rises, it is called convection.

The air in a room is constantly moving, in order to try and equalize the temperature in that space. Baseboard radiation heating systems use the principle of convection to help distribute warm air throughout a room, Figure 3-5. If warm air rises, cool air falls, forming a natural convection current of constantly moving air.

Modern furnaces utilize these basic principles of heat movement. To start, the combustion process releases heat energy, which is confined in a metal heat exchanger. The walls of the heat exchanger then become nearly as hot as the fire. Cooler air picks up this heat energy and becomes

Figure 3-5. Baseboard Radiation Using The Convection Principle

warmer as it passes along the outside surface of the heat exchanger. (An electric furnace eliminates this step by supplying heat energy directly to the air.) The heat exchanger design permits the air to pass over its entire surface, picking up a large amount of the heat energy available. Most of this fire to heat exchanger to air transfer is conduction, but some heat does pass due to radiation.

Due to convection and the force of the blower, air rises as it enters the space, then mixes with and warms the air in the space. The air not only warms the existing room air, it also replaces the air lost through the structure. Since the warm air rises, supply registers are almost always placed in the floor and under windows. This location allows the warm air to pick up and mix with the cold air falling down from windows or other cool surfaces. When the warm air mixes with the cold air coming down from the windows, it is called entrainment. This helps eliminate drafts from the house.

In short, in a ducted, forced warm air heating system, air is used as a heat transfer medium to move heat energy from its source (the furnace burner) to its destination (the living space) where it maintains predetermined comfort conditions.

FLUES AND CHIMNEYS

Flues and chimneys are vents used to remove the products of complete and incomplete combustion to outside the living area. While the products of complete combustion are not dangerous, these products are not desirable within the living space and should be vented to the outside atmosphere. On the other hand, incomplete combustion produces very harmful products, as mentioned earlier, and these products must be vented to the outside.

Proper venting of gaseous by-products is a very important part of any heating system. Houses with gas- or oil-fired appliances normally have a chimney running from the lowest floor to the roof. Houses without a chimney require a flue, and the same principles apply. In order to minimize flue length, the furnace should be located as close to the chimney as possible. A flue pipe connects the furnace to the chimney, Figure 3-6. When two or more appliances vent into the chimney, the one with the largest flue should be the lowest, Figure 3-7. An alternate method is shown in, Figure 3-8.

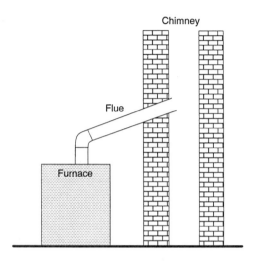

Figure 3-6. *Flue Pipe Connecting the Furnace to the Chimney*

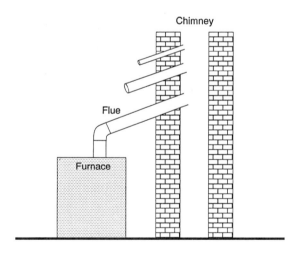

Figure 3-7. *Largest Flue Below Other Flues*

A chimney is simply a confined space leading upwards to the outside air. If the temperature of the air within the pipe is the same temperature as the air surrounding the pipe, there is no air movement. When the temperature of the air inside the pipe increases, it expands, becomes lighter, and rises. The warmer the air becomes, the faster it rises. When warm air flows upward in a steady fashion, it creates a good draft. As the warm air rises, it creates

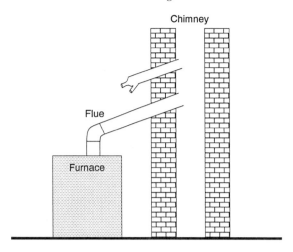

Figure 3-8. *Joining Two Flues*

a void which is filled by new air. This new air, when heated, also rises, continuing the flow for as long as there is a source of heat.

For a good draft to be maintained, the furnace must be on. This works out well, as the only time products of combustion need to be removed is during the combustion process.

Because the volume of flue gas a chimney removes depends on the height and diameter of the chimney, as well as the temperature of the gas, it is extremely important that the chimney be sized correctly. For example, a chimney that is too wide or too short can cause the gas to lose its velocity and subsequently stop the flow up the chimney. If the diameter is too small, the resistance (due to friction) reduces the draft. A very cold chimney exterior can hinder the draft and cause a downdraft until the inner surface heats up, causing gases to cool off and slow down before they reach the outside.

Condensation is another problem caused by a cold chimney. The combustion process produces a large amount of moisture. The flue then carries this moisture up and vents it to the outside. However, a large masonry or brick chimney does not warm up quickly, resulting in condensation in the chimney. Condensation causes problems, because flue gases have a normal amount of sulfur present. As such, when sulfur combines with water vapor, it produces sulfuric acid which erodes certain types of masonry and dissolves the mortar out of the chimney.

Installation of a liner with a surrounding insulating air space solves this condensation problem. Liners with overlapping joints are best because condensation or moisture drains all the way down to the bottom of the chimney. An oversized diverter at the furnace can also solve this problem. This diverter allows more warm air from the furnace room to be drawn in and up the chimney. If condensate forms in the flue pipe, insulating the pipe may help solve the problem.

GAS FURNACE VENTING

Twenty percent of the total heat produced by the furnace is used to heat the flue gases so that they are warm enough to produce a good draft. The house or living space uses the remaining 80 percent. For this reason, the input rating of gas is 80 percent of a furnace's heating efficiency. This is the principle a gas furnace uses to ensure efficiency and to ensure sufficient air is present for complete combustion to take place. Matching the burner temperature and the heat exchanger air volume produces a draft effect which allows the proper amount of combustion air into the compartment. This air is independent of the chimney draft, whose primary function is to eliminate the products of combustion.

RELIEF DEVICES

In a gas furnace, a relief device called a draft diverter or hood is built into the cabinet between the flue and the furnace, Figure 3-9. This is a prevention device that keeps the chimney draft from drawing excess combustion air into the furnace, thus lowering the temperature of the air flowing over the heat exchanger and reducing efficiency. The bottom of the diverter is open to the surrounding air, and the flue to the chimney is connected to the top or side of the diverter. In a gas furnace that is performing properly, the flue gases pass from the top of the heat exchanger, along the top of the diverter, then into the flue and up the chimney. Additional room air is also drawn in at the bottom and up the chimney, as the chimney draw is usually more than is necessary to remove just the flue gases.

SPILLAGE

There are several factors that can upset the chimney draft. These factors include: a restriction in the chimney which can reduce the volume of gas; a stoppage in the chimney which can keep gas from being vented outside; the outside temperature of the wind, which can cause a downdraft. If one of these factors occur and the flue or chimney completely stops up, the furnace can still operate. However, the flue gas must go somewhere, or else the back pressure will upset the combustion process in the furnace. The diverter compensates for this by allowing the gases to vent into the room. This is called spillage. In the case of complete combustion, this is not particularly critical, since the only effects are raised humidity and circulation of harmless carbon dioxide. If, however, incomplete combustion occurs, harmful elements such as carbon monoxide and aldehydes are circulated. These gases are extremely irritating, dangerous and potentially fatal. Consequently, whenever spillage is discovered, the cause must be found and corrected immediately.

Draft spillage is simple to diagnose. In most cases, gas odor and high humidity accompany this problem. If there are complaints of running eyes, headaches or dizziness, the service technician should suspect incomplete combustion and excess spillage. Testing for spillage is relatively easy. All that is needed is a lit match held about 1/2 inch below and in front of the diverter; the furnace must also be operating, Figure 3-10. If the diverter draws up the flame or smoke, it is working normally. If the flame or smoke blows back into the room, spillage exists and the cause must be found and repaired.

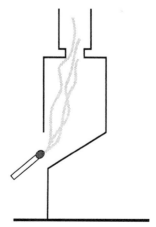

Figure 3-10. *Testing for Spillage*

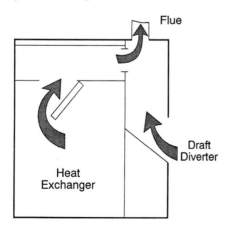

Figure 3-9. *Draft Diverter*

OIL FURNACE VENTING

An oil furnace comes with a combustion air blower. Therefore, the removal of flue gases is somewhat different than for a gas furnace. An oil furnace is not dependent upon the furnace chimney effect for combustion air, so no diverter is required. Instead, the oil furnace pipes directly into the chimney.

A barometric damper is installed in the flue pipe to provide draft control. The damper is hinged in the middle and has an adjustable weight, so it can swing freely, Figure 3-11. The damper swings open when the draft becomes excessive and draws room air into the flue. The weight is normally set at approximately 0.04 inches of draft using a water gage. Downdrafts do not normally upset combustion in an oil furnace, because the combustion air blower has enough power to overcome them.

Figure 3-11. Barometric Damper in Oil Furnace Flue. Courtesy, Field Controls Company.

Incomplete combustion in an oil furnace usually results in excessive carbon or soot, rather than carbon monoxide. Also, there is less water vapor produced than with a gas furnace, so combustion relief is not really necessary. However, a good flue and venting system is necessary for proper operation of all furnaces. Even if the service technician has no control of the original application, certain points on both the flue and chimney should be checked, as they can affect furnace operation.

EXAMINING FLUE TO CHIMNEY

When examining flue to chimney connection, the following points should be checked:

- To offer minimum resistance, the joints must be tight and positioned in the direction of flow. The fitting at the chimney must also be tight.
- The flue pipe connecting the furnace and flue must be the same size as the furnace collar through its entire length.
- The flue pipe should not extend beyond the inner face or liner of the chimney.

- To avoid friction losses and maintain flue gas temperature, horizontal flue pipes should be as short as possible and never more than 10 feet long.
- To avoid friction losses, the flue pipe should be as straight as possible.
- The pipe must have a pitch upwards of at least 1/4 inch/foot of length in horizontal runs.
- The pipe should have no obstructions.
- Insulating the pipe may help if there is poor draft or condensation in the pipe.
- After a heavy snow, flue pipe outlets should carefully be checked to ensure snow on the roof is not obstructing the flue.

EXAMINING THE CHIMNEY

When examining the chimney, the following points should be checked:

- To prevent downdrafts, the top of the chimney should be at least 2 feet above the highest point on the roof.
- The chimney top must be at least 3 feet higher than the point at which it passes through the roof.
- The coping should not restrict the opening.
- Loose building materials, displaced bricks or other debris should not clutter the chimney. This can be checked by lowering a flashlight down the chimney.
- Joists or other structural members should not protrude into chimney. This can be checked with a flashlight.
- The tiles and chimney face should show no signs of leakage. This can be checked by starting a smoking fire and observing if leakage occurs.
- The clean-out door should fit tightly.
- No opening should exist between flues using a common chimney.

Low or negative pressure at the furnace can also cause an improper draft. A downdraft can result if the pressure in the area is lower than atmospheric pressure. This can cause spillage or poor combustion.

COMBUSTION AIR IN FLUE

In order to work properly, the flue system requires an adequate supply of combustion air. The furnace 'starves' when any situation reduces the normal air supply obtained from infiltration and circulation. The most common causes of this condition are:

Insufficient Furnace Room Air. If air entry to the furnace is blocked or covered, or if air flow into the furnace room is restricted in any way, the furnace will starve for combustion air. Additional air openings, such as a louvered basement door, or outside air intakes usually solve this problem.

Furnace Blower Compartment Door Is Off. The compartment door should be kept on when the furnace is operating. This is because the suction of the blower can cause negative pressure in the immediate area, which can then unbalance the flue system and result in spillage.

Exhaust Fans or Vents Near the Furnace. Clothes dryer vents, kitchen exhaust fans or window-mounted exhausts located near the furnace all remove air from the immediate area. If this is the case, additional openings into the furnace area, or possibly outside air intakes are needed.

When performing checks or tests on the furnace, the service technician should test under normal operating conditions. This is to ensure an accurate diagnosis. This means that doors are on the furnace, windows are closed, basement door is shut, exhaust fans are running, etc. Even if an exhaust fan only runs intermittently, it should be turned on to give results under the most adverse conditions.

FURNACE CLASSIFICATIONS

Depending upon the type of house and its basic construction, furnaces can be used in a variety of ways. The furnace application is not tied to its energy source; therefore, the physical location of the furnace cabinet and its ductwork is the same for gas, oil and electric furnaces. For the most part, homeowners choose an energy source based on economic factors, the availability of a fuel in a certain area, and personal preference.

Regardless of energy source, the furnace cabinet houses the blower and heat exchanger, as well as the attendant controls necessary to operate them. The blower is always located between the return air inlet and the heat exchanger, and the filters are always located between the return air inlet and the blower. When an air conditioning system is installed along with the furnace, the air conditioning coil is located between the heat exchanger and the supply air plenum.

Furnaces are classified by the direction of air flow through the furnace, so there are upflow, downflow, reverseflow, counterflow, and horizontal-type furnaces.

UPFLOW FURNACE

In an upflow furnace, the return air enters the furnace at the bottom or side, flows up through the furnace, and into the supply ductwork, Figure 3-12. Normally, an upflow furnace is used in a basement. This way, the supply air runs can be put up high in the basement ceiling, directly underneath the floor of the living space.

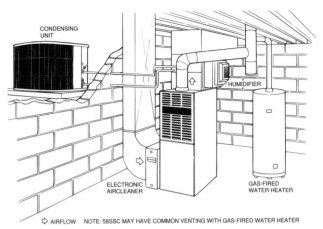

NOTE: 58SSC MAY HAVE COMMON VENTING WITH GAS-FIRED WATER HEATER

Figure 3-12. Upflow Furnace. Courtesy, Carrier Corporation, a Subsidiary of United Technologies Corporation.

DOWNFLOW, REVERSEFLOW AND COUNTERFLOW FURNACES

When the return air enters the top of the furnace and flows downward through the furnace, it falls into this category. Here, the supply air plenum and ductwork are located at the bottom. This type of furnace works best when put in a closet of a single-story house. This way, the furnace can still take in return air at the top and distribute the supply air through ductwork running under the floor. This type of furnace also works well in a house that does not have a basement but requires perimeter-type distribution, Figure 3-13.

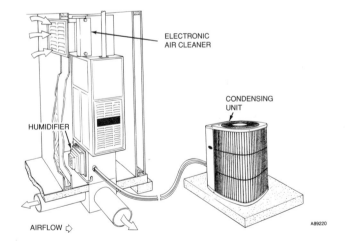

Figure 3-13. Downflow Furnace. Courtesy, Carrier Corporation, a Subsidiary of United Technologies Corporation.

HORIZONTAL-TYPE FURNACE

In this type of furnace, air flows horizontally through the furnace, entering at one end and discharging at the other end, Figure 3-14. This type of furnace works well in a house with a crawl space. The crawl space houses the furnace, and the ductwork runs under the floor, terminating under the windows.

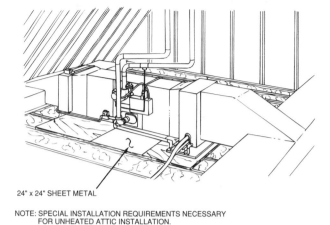

24" x 24" SHEET METAL

NOTE: SPECIAL INSTALLATION REQUIREMENTS NECESSARY
FOR UNHEATED ATTIC INSTALLATION.

Figure 3-14. *Horizontal-Type Furnace. Courtesy, Carrier Corporation, a Subsidiary of United Technologies Corporation*

Another way to use the horizontal-type furnace is to place it in an attic, with the supply air either ducted down through the wall and discharged low, or ducted directly to a high side wall, or through ceiling diffusers. This type of horizontal flow furnace, and/or combination heating and air conditioning system, can also be put outside. In this case, the supply air runs through the wall and into the house duct system. Mobile homes often use this method.

SUPPLY AND RETURN AIR

In a conventional system, the main trunk duct, often called an extended plenum, supplies air and runs lengthwise through the house. Smaller ducts, branching off the main duct, supply the various registers. These registers are usually found in the floor or on the wall.

The furnace receives returned air through a return air system that uses grilles located in central collecting areas, such as stairwells, or grilles cut into a sidewall or the floor. This air is then ducted back to the furnace. The simplest type of duct or return air system can be made by placing a piece of sheet metal across the floor joist, called a panned joist, Figure 3-15, and using this cavity to return the air to the furnace.

Many furnaces have a return air drop, which is a small cabinet designed to fit beside the furnace. Its purpose is to direct the return air from the top, where it is collected from the return air duct system or panned joist, to the bottom of the furnace. An extra cabinet may not be necessary, however, as many furnaces have two compartments: one for the heat exchanger, gas valve, burners, diverters, coil and electrical; and a matching cabinet for the blower, electrical make-up, humidifier and electronic air cleaner. This second compartment can also act as a return air drop.

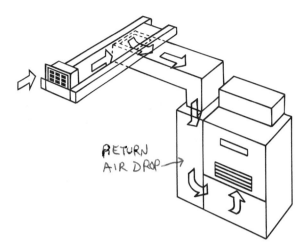

Figure 3-15. *Return Air System Using Pan Joist*

Most systems require some additional outside fresh air to replace air that is lost through the walls. In the case of gas or oil furnaces, extra fresh air is required to ensure a constant supply of combustion air in the basement. To receive this air, a fresh air intake is placed outside and connected into the return air system somewhere ahead of the furnace filters. This allows the fresh air to mix with the return air. This mixing raises the air temperature prior to the time it passes over the heat exchanger. The return air pipe is insulated in colder climates where condensation and frost are a problem. The return air pipe also has a manual damper which is designed to control the amount of fresh air allowed into the system.

A series of supply registers or diffusers supply air to individual rooms, Figure 3-16. These registers are placed on the floor, directly under the windows and about 18 inches from the outside wall, in colder climates. This placement allows drapes to close and prevents other obstructions close to the window from interfering with the flow of supply air. As stated previously, a process called entrainment then occurs; that is, the air flows upward from the registers, picks up any cool drafts coming off the windows, and mixes the warm air with the cool air. The result is an even temperature.

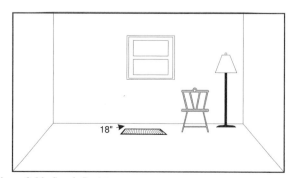

Figure 3-16. *Supply Register*

The mixed air moves around the room in a circular pattern, eventually returning to the floor. This movement allows warmed air to circulate freely throughout the space, eliminating the possibility of cold air falling to the floor and creating a draft. The mixed air then moves toward the return air register. As long as the furnace blower is running, this circulation continues. For this reason, it is recommended that the furnace blower be allowed to run constantly, regardless of whether or not the furnace burners are on.

When the furnace is on, the supply air is approximately 90 °F above room temperature. However, due to its velocity, the air feels cool to the touch when it enters a room through a register. As long as a person does not sit directly in the supply air stream, this will not cause any discomfort.

The size of the duct runs (calculated in the initial installation) controls the air. Each run contains a damper, which is located close to the furnace or take-off from the plenum. This damper controls the volume of air in each run. A damper at each register also controls air volume. These two dampers maintain balanced air volume in the system.

FURNACE SELECTION

The house construction, including whether or not a basement, crawl space or attic is available, normally determines which type of furnace is selected. The supply duct system largely depends upon the type of climate. In cold climates, there is more emphasis on heating than cooling, so the perimeter floor system is used. In warm climates, cooling is more important than heating, so a high side wall or high discharge is used. As hot air rises and cold air falls, either supply air system is a compromise between heating and cooling efficiencies.

REVIEW QUESTIONS

1. Describe kinetic energy.
2. What is the definition of temperature?
3. What is meant by Btu?
4. To achieve combustion, what three elements must be present? FUEL, AIR, HEAT
5. What are three fuels generally used in heating?
6. Name the basic elements in all fuels. HC
7. What is complete combustion?
8. What is incomplete combustion?
9. Name the products derived from incomplete combustion. CO, KETONES, ALDEHYDES, SOOT
10. Describe the difference between a yellow flame and a blue flame. BLUE FLAME HOTTER, COMPLETE
11. What is primary air? PREMIXED
12. What are three methods of heat transfer?
13. What is the purpose of a chimney and a flue?
14. What is a draft diverter?
15. Describe a simple test for spillage. LIT MATCH
16. In an oil furnace, how is draft control accomplished?
17. What three situations cause a furnace to starve?
18. How are furnaces classified?
19. In what way is the air delivered to the individual rooms?
20. How do you control the amount of air delivered to a room?

Chapter 4

Basic Heating Components

Regardless of the energy source, heating components such as blowers, filters and cabinets are common to all types of furnaces. This chapter covers these components, as well as components that do vary according to the type of furnace. No matter which components are used, however, the basic principles remain the same.

FURNACE CABINETS

An airtight sheet metal cabinet houses the basic heating components. Figure 4-1 illustrates one of the many different furnace cabinets available.

Figure 4-1. *Furnace Cabinet. Courtesy, Carrier Corporation, a Subsidiary of United Technologies Corporation.*

A collar is provided for the attachment of the supply air plenum (located on the top of an upflow furnace). Knockout panels on each side indicate where the return opening is left intact. These panels are necessary, because the manufacturer does not know which side will be used for return air at the time the furnace is built. The installer can then cut the appropriate opening on the job site. Furnaces may also have a knockout panel located on the bottom. This is necessary for mobile homes or apartments which require the return air to come in below the furnace. Electrical connections and fuel piping also require knockouts.

Cabinet size varies between manufacturers, because the size is dependent upon the space required for the heat exchangers and burners, as well as space to move the required amount of air. In certain cases where height is the major consideration (as with a low-boy furnace designed for basements with low ceilings), two cabinets can be placed back-to-back. Of course if height is not a factor, all components are combined in one cabinet.

All furnaces require access panels or doors which allow the service technician to get to the burners, blower compartment and other operating components. These doors may snap on or off, or have latching handles. The doors may have louvers or just an open area, in order to allow combustion and vent air to enter. As no combustion takes place in electric furnaces, this type of furnace does not require provisions for combustion air or air for venting. Consequently, electric furnaces can have solid panels or doors. Electrical make-up boxes, for line and low voltage, are usually located in the blower compartment. An oval flue connects a gas or oil furnace to the top of the cabinet. An electric furnace does not require this.

Manufacturers often bolt each corner of the furnace to a wooden shipping frame. After removing the shipping frame, these bolts can be re-inserted and used as leveling bolts during installation.

FILTERS

Heating and air conditioning units require filters. These filters are a very important part of the system, as they remove particles from the air stream which could potentially damage the furnace. A dirty filter reduces efficiency and can give the appearance of a serious problem. Often, a service technician finds a customer's complaint is due to a dirty filter.

While filters differ in make and efficiency, the blower compartment always contains some type of filter between the return air side of the system and the blower, Figure 4-2. A slab filter is the most common filter. Slab filters can either be throwaway or permanent, and they come in a variety of standard sizes. They are usually 1 inch thick.

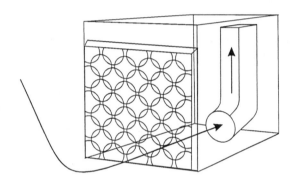

Figure 4-2. *Filter in Return Air Ahead of Blower*

A throwaway slab filter consists of a Fiberglass filter held in place by a cardboard frame. The front and back of the Fiberglass are covered by very thin metal, with large holes cut into the metal. When this type of filter gets dirty, it is removed from the furnace and thrown away, and replaced with a new one. Filters can be mounted in the furnace horizontally, vertically or in combination, Figure 4-3. They are located ahead of the blower and near the return air inlet. A rack or a rail provides easy insertion and removal.

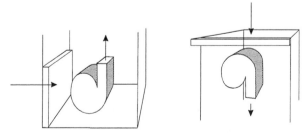

Figure 4-3. *Filters Mounted Horizontally or Vertically*

A permanent slab filter can be constructed of polyurethane or a metal fiber mesh. A wire frame holds the polyurethane filter and clips hold the filter in place. The

material in a metal fiber filter cannot be removed from its frame. However, either type of material can be removed from the furnace and cleaned, and then reinstalled.

To change a polyurethane filter, perform the following steps:

1. Remove the filter from the frame.
2. Using a special cleaner, wash the filter in water.
3. Squeeze the water from the filter.
4. Replace the filter in the frame.
5. Oil the surface of the filter and reinstall in the furnace.

BLOWERS

One of the most important components in a heating or cooling system is the prime air mover, called the blower. The blower has a diameter wheel or scroll, of approximately 8 to 10 inches, mounted in a metal enclosure (housing) with an opening offset to one side. When the blower rotates, air is taken in at the side and then pushed out the opening at the top under pressure. There are two basic types of blowers: the belt drive blower and the direct drive blower.

BELT DRIVE BLOWER

The wheel or blower is mounted on a shaft set in bearings on each side in a belt drive blower. A platform or cradle holds the large grooved pulley. In order to minimize noise and vibration, rubber is used to cushion both the motor and mounting hardware. Due to the motor mounting frame design, the motor can be moved left and right or up and down. This is necessary so the pulley on the motor can be exactly aligned with the blower pulley. Two adjustable tension metal straps, one on either side, usually attach the motor to the cradle. Figure 4-4 illustrates a belt drive blower.

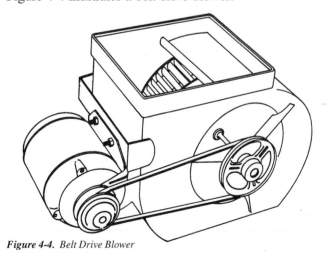

Figure 4-4. *Belt Drive Blower*

DIRECT DRIVE BLOWER

When the blower motor is mounted inside the blower scroll, and the drive shaft is connected directly to the wheel, it is called a direct drive blower, Figure 4-5. With this type of blower, the blower runs at the same speed as the motor. A direct drive blower is closely matched to the furnace and its intended application.

Figure 4-5. Direct Drive Blower. Courtesy, Carrier Corporation, a Subsidiary of United Technologies Corporation.

BLOWER MAINTENANCE

The procedures outlined in the following paragraphs pertain to both belt drive and direct drive blowers. While specific procedures may vary somewhat, the basic steps are valid for both blowers.

Cracks, splits, ragged edges, uneven wear, or severe discoloration all indicate a new blower belt is necessary. As a worn or stretched belt can cause many problems in adjustments or wear to the equipment, all belts should be replaced if there is any question as to their ability to perform properly. The correct size and number is listed on each belt to ensure the proper belt is replaced. A belt that is too large, too small, too narrow or too wide eventually causes problems that are easily avoided by proper sizing at the outset.

To Remove Blower Belts:

1. Turn the furnace off at the unit as well as the main disconnect.
2. Loosen motor mounting screw.
3. Release belt tension by moving the blower motor up.
4. Remove the belt from both pulleys.

To Check the Blower and Motor Bearings (Belt Off):

1. Check the setscrew on the blower pulley for tightness.
2. Try to move the blower shaft up and down, back and forth. The shaft should not move in any direction. If it moves, the bearings must be replaced.
3. If, when pushing the blower shaft in and out there is greater than 1/8-inch play, loosen the setscrew on the thrust collar and move the collar closer to the bearing. Then, retighten the setscrew.
4. Try to move the motor pulley up and down, back and forth. If the pulley moves, the motor bearings should be replaced.
5. Push the motor shaft in and out. It should have about 1/8-inch play.

To Check Pulley and Drive Alignment:

1. Place a straight edge along the side of the pulleys, or place a straight 1/4-inch rod in the grooves of both pulleys. If the rod is not absolutely straight, loosen shaft setscrew on motor pulley and realign.
2. Tighten setscrews and recheck with rod or straight edge.
3. Check all setscrews and motor bracket screws for tightness.

To Lubricate Motor and Blower Bearings:

1. If the bearings do not have grease cups or fittings, presume they are permanently lubricated fittings that do not require servicing.
2. Turn down the bearings with grease cups about one turn/year. Refill the cups once they are turned all the way down. Always check the manufacturer's instructions for the type of grease to use.
3. If bearings do have grease fittings or plugs, remove the plug from the bearing, add a 5-lb relief fitting, and lubricate with a No. 2 consistency neutral mineral grease.
4. Check manufacturer's instructions before lubricating motor bearings. If there are no instructions and the motor has oil cups or holes, add a few drops of SAE No. 10 nondetergent oil twice a year.

CAUTION: Do not overlubricate, as excessive oil can damage the starting switch or spill over and damage the belt.

To Check Belt Tension:

1. Press down on belt with finger.
2. The belt should give (flex) 1 inch when moderate pressure is applied to the top of the belt, halfway between the pulleys. If not, loosen the mounting screw bracket and readjust and retighten the screw.

PULLEYS

Most belt drive blowers have an adjustable motor pulley which is used to change the blower speed. During initial heating installation, this pulley is sometimes needed to adjust air volume. It is also sometimes used to increase the air volume when the system changes over to cooling.

Adjustable pulleys have an Allen setscrew in the outer flange which, when loosened, allows the flange to be screwed in or out, Figure 4-6. In order for an adjustment to be made every half turn, some pulleys have two flat areas in the threads, separated by 180 degrees. Others have only one flat area, in which case, speed adjustments can only be made for one complete turn. The setscrew should be located over the flat area of the shaft, since any other location can strip the threads. Rotating the outer flange clockwise increases the inside diameter of the pulley, making the blower run faster. Rotating the flange counterclockwise decreases the inside diameter of the pulley, reducing blower speed.

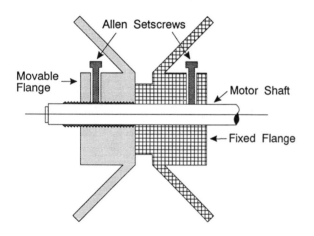

Figure 4-6. *Adjustable Pulley*

Pulleys, motors and drives are matched to the furnace air requirements. As such, the adjustment range is usually capable of supplying the necessary air for most normal applications. If more air is required, it is possible to change the entire pulley to the next larger size. However, changing pulleys can be risky, because if a great deal of additional air is needed, a larger blower motor may be required. If the motor is too small, it will overload and be unable to supply the air required.

BLOWER MOTORS

There are several different types of motors used to drive the furnace blower. The running and starting loads imposed by the application determine the motor. These motors include split-phase, capacitor-start, shaded-pole, and permanent-split capacitor motors.

SPLIT-PHASE MOTOR

This motor develops starting torque by having the start and run windings out of phase with each other, Figure 4-7. Both windings are in the circuit when the motor starts. However, when the motor comes up to speed, a centrifugal switch opens and disconnects the start winding from the circuit. Because this motor has relatively low torque, it can only be used on applications with light starting loads. It is used on belt drive applications 1/3 hp in size and smaller. This is usually a four-pole motor which operates at 1,750 rpm.

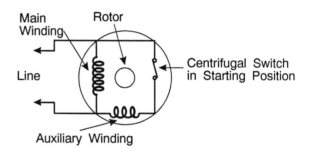

Figure 4-7. *Split-Phase Motor Schematic*

CAPACITOR-START MOTOR

This motor consists of a starting capacitor mounted on the motor and internally wired into the start winding, Figure 4-8. This additional assistance to the start winding gives the motor considerably more starting torque than the split-phase motor. Therefore, a capacitor-start motor can start against greater loads than a split-phase motor. When the motor comes up to speed, a centrifugal switch opens, taking the start winding out of the circuit. A capacitor-start motor is used on belt drive applications 1/3 hp in size and larger.

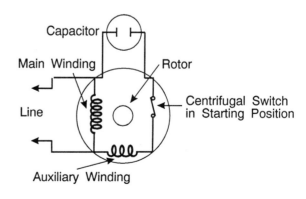

Figure 4-8. *Schematic of a Capacitor- Start Motor*

SHADED-POLE MOTOR

Loops of copper wound around each pole give the shaded-pole motor its starting torque. This configuration places the windings enough out of phase to start under relatively light loads. This motor has the lowest starting torque of any motor described in this section and is used basically for smaller residential furnaces. The shaded-pole motor is used on direct drive blowers in 1/10 to 1/6 hp sizes. It is a six-pole motor operating at 1,050 rpm.

PERMANENT-SPLIT CAPACITOR (PSC)

This motor is a single-phase induction motor designed to use a single fixed capacitor for both starting and running duty, Figure 4-9. The capacitor has two functions: to help provide the extra starting torque required when the motor starts, and to help improve the efficiency of the motor while it is running. This capacitor is permanently connected in the circuit and splits the phase of the current in the auxiliary (starting or phase) winding with respect to the main (run) winding, hence the name permanent split capacitor. Consequently, a PSC motor has more starting torque and better running efficiency than a shaded-pole motor, because the capacitor stays in the circuit. A PSC motor is a six-pole motor operating at 1,050 rpm and is used on direct drive applications from 1/8 to 3/4 hp.

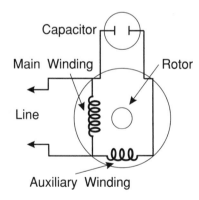

Figure 4-9. Schematic of a Permanent Split-Capacitor Motor

AMPERAGE DRAW

The amperage drawn by any motor is an indication of its operating condition. The service technician takes and records this reading during every service call. An overloaded motor is the problem when the amperage is greater than the nameplate rating. Restrictions in air movement, often due to dirty filters, can cause this problem. If the amperage is less than the nameplate rating, the motor is not working to full capacity.

MOTOR SPEED CONTROL

Blower speeds must sometimes be changed, in which case, the amount of air delivered to the space also changes. Switching from heating to cooling is the most common situation for changing speeds.

A speed control, or series reactor, can electrically change the speed and vary the voltage. As the speed of any motor is a function of voltage and load, varying the voltage while keeping a fairly constant load changes the speed of the motor.

A series reactor consists of an iron core wrapped with many turns of wire, similar to one side of a transformer. Usually three or four wire leads tap into the reactor at various points, Figure 4-10. The motor runs at high speed if it is connected to the first lead of a speed control, because in this position, it is supplied with full line voltage. When connected to the second lead, the motor runs at a slower speed, because pushing the current through the first group of turns in the reactor uses up some of the voltage. As a result, the motor receives less voltage and runs more slowly. The motor runs at its slowest speed when connected to the third lead, as the reactor uses up the additional voltage.

The reactors should not allow a motor to run below 550 rpm. At this point, the lubrication of the motor bearings reduces to the point where the bearings fail.

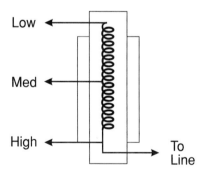

Figure 4-10. Series Reactor

The taps are usually color-coded. The furnace manufacturer then provides a wiring diagram identifying these color-coded taps. To prevent a short circuit, unused taps should be taped separately and located out of the way. A voltage reading should be taken across any unidentified taps in order to determine the speed; the higher the voltage, the greater the speed. The series reactor can be used with voltage-sensitive direct drive motors. The series reactor can either be supplied on the original installation or added to existing installations which have shaded-pole or split-phase motors.

TAP-WOUND MOTORS

A tap-wound motor, available with three, four, or five speed choices, is another way to provide multispeed motors for direct drive applications, Figure 4-11. Both shaded-pole and PSC motors are available with these features. By tapping into the motor windings and bringing these leads outside, many speeds can easily be achieved. Just as with a speed control, the greater the length of winding used, the slower the motor speed.

Figure 4-11. Tap-Wound, Three-Speed Motor. Courtesy , Carrier Corporation, a Subsidiary of United Technologies Corporation.

All leads are color-coded according to the speed they produce. Many manufacturers color-code as follows:

Color	5 Speed	4 Speed	3 Speed
Orange	Common	Common	Common
Purple	Capacitor	Capacitor	Capacitor
Black	High	High	High
Brown	Med. High	Med. High	
Blue	Medium		Medium
Yellow	Med. Lo	Med. Lo	
Red	Low	Low	Low

NOTE: Color-coding may vary, depending upon the manufacturer. Before connecting, the wiring diagram should be checked.

As in a speed control, the unused leads should be taped separately to avoid shorts. Some leads have spade terminals which plug into a terminal box inside the furnace housing. This box may have insulated dummy terminals for the unused leads.

The cooling relay may be capable of selecting one of two motor speeds. This allows automatic selection, at the thermostat, of the proper speed for either heating or cooling. One initial speed selection is all that is required to have one speed for cooling and another (slower) speed for heating.

FAN CONTROL

All types of forced air furnaces (upflow, downflow, horizontal) contain a fan control. The fan control is located about one-third of the way down in the heat exchanger. Figure 4-12 shows a fan control, which, in most newer furnaces, is accomplished by the microprocessor control board.

The fan control has two major functions: (1) when heat is needed, the fan control turns the blower on; it then turns the blower off when the desired heat level is reached, and (2) the fan control provides a time delay before turning the blower on and turning it off. This time delay allows the furnace to warm up before it delivers the heat, thus avoiding drafts and ensuring air of the proper temperature is delivered to the living space. When the furnace shuts off, the time delay allows additional air movement in order for the heat exchanger to cool down.

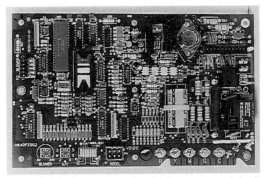

Figure 4-12. Fan Control. Courtesy, Carrier Corporation, a Subsidiary of United Technologies Corporation.

The fan control contains a set of normally-open contacts and is in the line voltage circuit, in series with the blower motor. The bimetallic element in the fan control inserts into the heat exchanger. This way, the element is in the exiting air stream and can directly register the temperature of the supply air.

There are two types of fan controls. In one type, both on and off temperatures are adjustable. In the other type, only the off temperature is adjustable, with a fixed temperature differential for the on cycle.

The set point for turning the blower on is usually between 90 and 110 °F. This allows the blower to turn on quickly and direct the heated air to the space rather than to the flue. The differential (the difference between the point where the blower comes on and the point when the blower turns off) is usually between 15 and 35 °F (25 °F in a fixed differential fan control). For example, if the blower turns on when the air leaving the heat exchanger reaches 105 °F, the blower will shut off when the air cools down to about 80 °F. These set points can be changed to suit the customer; however, the furnace operation is more efficient when the on set point is lower. It is possible to set the fan control below the normal return air temperature in case a thermostat does not have a fan switch for continuous blower operation. This keeps the blower operating even when the burners are off.

Fan control calibration can be checked by inserting a thermometer in the furnace plenum and noting the temperatures at which the furnace turns on and off. While this reading does give reasonable indication, it is not completely accurate, due to the lag in the thermometer reading.

DOWNFLOW AND HORIZONTAL UNITS

Due to the fact that heat rises, the fan control in an upflow furnace responds correctly in any circumstance. This is not the case in a downflow furnace, where the burners can come on, warm up the heat exchanger and raise the temperature of the fan bimetal control to the make point. As a result, when the blower comes on, cooler return air is forced over the fan control. This cooler air can reduce the temperature to the break point, thus stopping the blower and possibly causing undesirable cycling of the blower under perfectly normal conditions.

It is a good idea to keep the blower on long enough for the heat exchanger to cool down once the burners turn off. In a downflow furnace, the return air is at a temperature below the setting of the fan control. This air 'fools' the control into thinking the heat exchanger is now cool. As a result, the blower turns off too soon. Once the blower is off, the residual heat in the heat exchanger rises, closing the fan switch and bringing the blower back on.

To avoid this problem, downflow furnaces use a special fan control. This control includes an electric resistance, or warp, switch wired in parallel with the control circuit. When heat is needed, the warp switch heats up before the bimetallic element in the furnace and closes the fan control switch, starting the blower. This takes about 60 seconds and can occur without the burner coming on at all. As the supply air comes up to temperature, the bimetal in the furnace, acts through a push rod and holds the contacts in. When the thermostat is satisfied, the circuit opens, and the resistance cools off quickly. The residual heat in the furnace causes the furnace bimetal to keep the blower on for about 1 1/2 minutes before breaking the fan control contacts.

LIMIT CONTROLS

Furnaces are equipped with a safety device called a limit control, a component designed to prevent the heat exchanger from overheating. Overheating can occur if the blower motor fails or if the air passing across the heat exchanger is restricted in any way. If a heat exchanger overheats for any period of time, holes can develop, leaking the products of combustion into the supply air. This is a health and fire hazard. A limit control protects against such problems.

A limit control has a set of normally-closed contacts and is usually placed in the low voltage circuit for gas furnaces and in the line voltage circuit for oil furnaces. It can be mounted separately or physically combined with the fan control. A bimetallic element placed in the air stream allows the limit control to sense the temperature passing over the heat exchanger.

Limit controls are usually nonadjustable and have a factory-set break point of 200 °F. When the limit opens, it breaks the control circuit to the thermostat, shutting off the furnace. A limit control automatically resets when the furnace cools to about 175 °F, so there is a built-in differential of about 25 °F between the make and break points.

Gas and oil downflow and horizontal furnaces require a secondary limit to ensure accurate sensing of all temperature conditions. The primary limit control, located in the lower part of the cabinet, monitors for overfire or restriction of air over the heat exchanger. The secondary limit control is wired in series and located above the heat exchanger. If the blower fails, the secondary limit will sense heat rising from the heat exchanger and open the contacts. On many oil furnaces, the secondary is a SPDT switch, one leg of which bypasses the fan control and tries to bring on the blower to cool off the heat exchanger, and at the same time, break the control circuit, shutting off the burner. Figure 4-13 shows a fan and limit control.

NOTE: These controls are safety devices used on gas and oil furnaces only; they do not apply to electric furnaces. The limit for an electric furnace is unique and will be discussed in the chapter Electric Heating.

Figure 4-13. Fan and Limit Control. Courtesy, White-Rodgers Division, Emerson Electric Company.

REVIEW QUESTIONS

1. What is a filter?
2. Where is the filter located? RETURN SIDE OF BLOWER
3. What are the two basic types of filters?
4. What are two basic types of blowers? BELT; DIRECT DRIVE
5. Describe how to check blower and motor bearings. ~WIGGLE
6. Describe how to check pulley and drive alignment. ~STRAIGHTEDGE
7. How is belt tension checked? ~ STRAIGHTEDGE
8. How is the air from the blower changed in a belt drive blower? ADJUST THE MOTOR PULLEY
9. What are the four different types of blower motors? PSC, SHADED POLE, CSIR, SPLIT PHASE
10. What is a speed control? ~ MULTISPEED GIZMO
11. Name the applications in which split-phase motors are used. LIGHT TORQUE SETUPS
12. What hp are capacitor-start motors on belt drive applications? 1/3 HP + BETTER
13. What is the difference between a shaded-pole motor and a PSC motor? PSC HAS A RUN CAP
14. What is a tap-wound motor? MULTISPEED TAPS ON MOTOR COIL
15. How does a tap-wound motor change speeds? CHANGE WIRES
16. What is a fan control? THERMOSTATIC OR TIME DELAY SWITCH
17. What is meant by normal differential in a fan control? OFF TEMP — ON TEMP
18. What is a limit control? ~ H/E SAFETY
19. Which circuit is a limit control on? LOW VOLTAGE, GAS VALVE.

Chapter **5**
Gas Heating

GAS FURNACE

The gas furnace is a heat-producing and heat-distributing system, Figure 5-1. The heat-producing part consist of heat exchangers, burners, and a venting system. This venting system allows flue gases to exit into the atmosphere.

The heat-distributing part of a gas furnace consist of a blower that moves the heated air into the ductwork. This heated air moves through the ductwork into the areas needing heat.

Figure 5-1. Gas Furnace. Courtesy Carrier Corporation, a Subsidiary of United Technologies Corporation.

HEAT EXCHANGERS

Gas furnace heat exchangers, often called clamshells, are constructed of two pieces of metal stamped into forms of left and right halves. These two pieces are then welded together along a common vertical seam. Cold-rolled steel or aluminized steel is often used to make the heat exchangers, and a special ceramic coating is often applied in order to resist rust and acid deterioration.

As the heat exchanger has many important functions, its shape is complex. The heat exchanger shape also depends on its application. Heat exchangers must provide room for the burners and allow them to supply steady, even heat over a maximum surface area inside the heat exchanger. In order to maintain the velocity of the air as it rises, many heat exchangers contain a reduced cross-sectional area. The upper portion of a heat exchanger may contain baffles, or it may be shaped in a zigzag fashion. These configurations help transfer the maximum amount of heat generated by the burners to the walls of the heat exchanger. Some heat exchangers may also have indentations, which give additional structural strength to the walls. This helps to keep cracking and popping noises from occurring when the metal expands and contracts. There is an exit at the top of the heat exchanger which allows the products of combustion to pass into the diverter and flue.

For efficient heat transfer to occur, the heat exchanger must be shaped so that the air passing over its outside surface picks up the maximum amount of heat. A sufficient quantity of air must pass over all the surfaces to remove heat and prevent hot spots from forming on the heat exchanger. These hot spots eventually burn through the heat exchanger if there is insufficient air.

Each heat exchanger has an output of between 20,000 and 50,000 Btu, depending upon its effectiveness in completing total heat exchange. To increase the output rating of the furnace, more than one heat exchanger may be used. For example, a unit with two heat exchangers is rated at 40,000 to 100,000 Btu, whereas a unit with three heat exchangers is rated at 60,000 to 150,000 Btu.

TESTING FOR LEAKS

One way of checking for leaks in a residential furnace combustion chamber is to spray a common table salt solution into the combustion chamber. The heated air is then tested with a small butane torch. If the salt solution mixes with the heated air, the color of the butane flame turns from blue to yellow, indicating a leak in the heat exchanger. This quick and simple technique works well for gas and oil furnaces.

BURNERS

Each heat exchanger in the furnace has an individual gas burner, Figure 5-2, so the number of heat exchangers determines the number of burners. The gas burner fits into an opening located on the bottom of the heat exchanger. The number of heat exchangers varies, depending on the Btu output of the furnace.

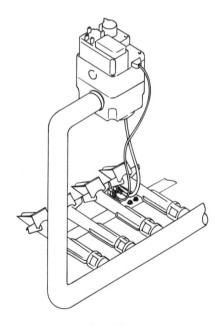

Figure 5-2. Complete Burner Assembly

Gas combustion occurs at the burners, and combustion requires primary and secondary air. An adjustable shutter or slide controls the quantity of primary air to the burner.

PRIMARY AND SECONDARY AIR

As discussed previously, primary air is brought into the burner tube and mixed with the gas prior to ignition. Secondary air is drawn from outside of the burner and joins with the burning gases at or near the burner ports, Figure 5-3. In a well-designed burner, the secondary air reaches the burner ports at an even rate and provides a good gas-air mixture. After the heat exchanger becomes hot, secondary air is brought to the burner by the draft created in the heat exchanger.

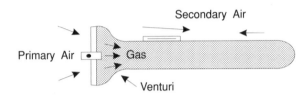

Figure 5-3. Supplying Primary and Secondary Air to the Burner

Due to the suction created by the gas flow into the tube, primary air instantly enters the burner when gas begins to flow. Increasing the amount of primary air that is entrained with the gas reduces the dependence on secondary air and improves the performance of the burner.

Insufficient primary air in a burner can cause the flame to smother when it comes on in a cold heater. When a flame smothers, it lifts off the burner and blows out the front, often extinguishing the pilot. Too much primary air can produce loud popping noises when the burners turn off. This occurs because a slight draft still exists after the burner turns off. This draft then pulls additional air into the burner, mixes the air with the remaining gas, and, as a result, the burner ignites, causing noise and extinguishing the pilot.

ORIFICES

The size of the gas orifice and the manifold pressure determine the amount of gas that flows to the burner. A small nozzle contains the gas orifice that controls gas flow. The larger the orifice, the more gas will flow. This nozzle attaches to the gas manifold and fits into the center of the burner.

The size of the gas orifice is determined by the specific gravity of the gas (usually about 0.6), the furnace input per orifice (about 25,000 Btu), and the heat value of the gas (about 1,000 Btu/ft^3 for natural gas). These factors determine whether an orifice size must be increased or decreased. Many manufacturers use data tables to determine the new orifice size. Orifice sizes are given as drill numbers; the smaller the drill number, the larger the orifice hole and the greater the Btu gas input. For example, a natural gas furnace using 1,000 Btu/ft³ gas, with a specific gravity of 0.6 and a 25,000 Btu input burner, uses a No. 38 drill. A 30,000 Btu input burner requires a No. 35 drill.

Orifices must be changed when the installation occurs 2,000 feet or more above sea level. Regardless of the gas being used, input ratings should be reduced by 4 percent for each 1,000 feet above sea level.

BURNER DESIGNS

Figure 5-4 illustrates several basic burner designs. The most common design is an elongated steel tube made of cast iron. This particular design has a narrow portion, called a throat or venturi, close to the entrance of the gas, and a wider mouth (for primary air) to which the gas supply is attached. An orifice in the gas manifold supplies gas to the end of the tube, and the gas then travels to the end of the burner. The velocity of the gas increases as it enters the venturi, which then helps pull in the primary air. The amount of primary air is controlled by either a shutter at the end of the burner or a rotating sleeve close to the throat. To reach the combustion ports, the gas-air mixture travels the entire length of the tube, which mixes it with the secondary air.

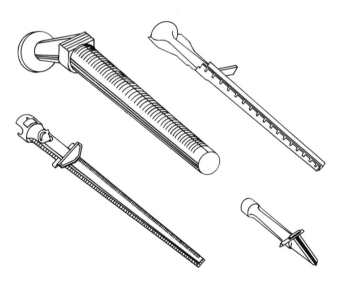

Figure 5-4. Various Burner Designs

Burner ports can be formed at the top of the burner tube by either an insert or an additional metal strip of stainless steel between the open space. These strips can be shaped either as ribbons or slotted crossways. Spot welds or rivets hold the insert in place, so the predetermined port openings are constantly maintained. Some designs have holes or slots cut directly into the burner.

CROSSOVER IGNITERS

When several burners are present, they must ignite quietly and simultaneously without delay, even when low gas pressure conditions exist. A standing pilot located near the center of the furnace causes ignition. Attached to each burner is a crossover igniter which contains an open slot. When heat is needed, each burner supplies some of the gas-air mixture to the crossover. The pilot ignites the gas, which turns all the burners on at the same time. Some crossover igniters include a small baffle or air scoop to divert part of the gas-air mixture down toward the main burner.

In the heat exchanger, burners are held in place by a small slotted piece of metal located near the crossover igniter. The burners can also be held in place by fitting them into a contour at the other end of the heat exchanger.

BURNER MAINTENANCE

A burner flame should be of a fairly soft blue color with no yellow present. If a yellow flame is observed (indicating unburned carbon), the primary air must be increased by opening the shutter or sleeve. To maintain the set position, the shutter or sleeve has a small setscrew. The air opening increases when this screw is loosened. The shutter is usually set to the point where a yellow flame appears, then adjusted just to the point where the yellow flame disappears. At this point, the shutter should be set and the setscrew tightened.

Approximately 50 percent primary air must be present in order to produce the required blue flame. As a completely closed shutter cuts off the source of primary air, the shutter should never be closed down completely.

PILOT BURNERS

A pilot burner, which is just a small gas burner, ignites the main burners. A piece of tubing, which feeds a small amount of gas to the pilot burner, connects the pilot burner to the main gas valve. The pilot burner is slotted, which allows the flame to extend out several inches in two directions, and mounted so that one of the flames is directed to the crossover igniter for the main burners. Thus, when gas is supplied to the main burners, they all light simultaneously. A slot or drilled hole in the burner head supplies primary air to the pilot burner.

A thermocouple is connected to the pilot burner assembly. A thermocouple is an electric generator that determines whether or not the pilot flame is lit. This is important to know, because when no pilot is present to ignite the gas as it enters the main burner, the gas collects in the furnace and surrounding area. This creates a very dangerous condition, because the gas collection can potentially explode. To prevent this from occurring, the pilot flame is directed to the top of the thermocouple. The current generated in the thermocouple indicates to the gas valve that the pilot is lit. Thermocouples are discussed in more detail later in this chapter.

LIGHTING THE PILOT BURNER

The nameplate found on every furnace contains details on how to light the pilot burner. The following steps are typically included:

1. Set the room thermostat either below the point where it calls for heat, or pull the furnace disconnect switch so that the furnace cannot come on when the pilot is lit.
2. Turn the manual valve on the gas valve to its pilot position; depress it and hold it down, Figure 5-5.

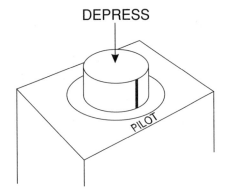

Figure 5-5. Depress Manual Valve Before Lighting Pilot

3. Hold a match to the gas outlet of the pilot, and depress the manual valve for about 60 seconds to allow the thermocouple to heat up. After this time period, rotate the manual gas valve to its on position, and release.

Some gas valves have a plunger, usually colored red, which must be pushed in to light the pilot. The procedure for the plunger is the same as for a manual valve.

The manual valve can be turned back to pilot, should seasonal or temporary shutoff be desired. Many valves contain a safety circuit which, when the valve is turned to the off position, does not allow the valve to be turned back to pilot for three to five minutes. The valve should not be forced.

PILOT BURNER MAINTENANCE

If burners become dirty, they need to be removed and cleaned. This procedure is outlined below.

1. Shut off the main gas supply and remove the screws that hold the gas manifold in place.
2. At the gas valve, remove the pilot gas line and thermocouple leads to the gas valve.
3. Rotate the manifold to remove the burners from the heat exchanger.
4. If needed, clean burner ports with a wire brush. If they are very dirty, a small piece of sheet metal, which fits into the burner ports, can be used to clear each port.
5. Clean the inside of the burners with a stiff, bottle-type cleaning brush.
6. After cleaning, replace the burners in the heat exchanger, taking care to relocate the pilot and thermocouple in their original locations.
7. Make sure the crossover igniters are lined up properly.
8. Check burner operation and ignition after reinstallation.

PILOT FLAME

It is important to observe the pilot flame. The flame should be a soft blue color and surround the top 3/8 to 1/2 inch of the thermocouple tips, Figure 5-6. If this is not the case, the problem may be due to improper location of the pilot flame or thermocouple, a dirty orifice in the pilot, or low gas pressure. The gas pressure to the pilot flame can be adjusted by turning a screw on the gas valve or, if there is a separate pressure regulator, it can be adjusted at this point.

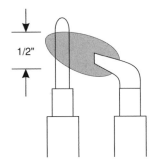

Figure 5-6. Thermocouple Surrounded by Pilot Flame

The following types of flames can signal trouble:

Lazy yellow flame. This indicates a dirty pilot burner when the combustion chamber is cool and the gas pressure is normal. However, if the pilot burner is clean and the combustion chamber is hot, the lazy yellow flame in this instance indicates high ambient temperatures.

Lazy blue flame. This flame waves about and has no well-defined shape, indicating burning at the orifice. A whistling sound, characteristic of burning at the orifice, also signals this condition.

Very small flame. If this flame occurs when valves are wide open and pressure is normal, a clogged orifice is indicated. However, a pilot burner filled with carbon produced by burning at the orifice can also cause this type of flame. While orifice clogging is usual with all gases, carbon build-up is unusual for manufactured gas. A dirty filter or a too-small filter in the pilot line may cause a very small flame. A clogged pilot filter, a too-small manifold, or insufficient pressure may be indicated if the pilot flame becomes very small once the main burner comes on. Enlarging the pilot orifice may relieve the problem.

Lifting or blowing flame. High pressure, a too-large orifice, or sometimes both, can cause this condition. This type of flame is noisy and sometimes blows itself out. Reducing the pressure to normal range and using the proper type and size of orifice inlet fitting corrects this type of flame. A noisy, roaring flame may result from placing the pilot in a strong, onrushing air stream. Shielding the pilot from this air stream will correct the trouble.

Hard, sharp flames. This condition is not usual for manufactured, butane-air and propane-air gases. In this case, the flame is noisy and has a tendency to blow itself out; or, it backfires and burns at the orifice. A dual-orifice inlet fitting of the proper type and size corrects this problem. Increasing the size of the orifice and decreasing the size of the primary air hole provides a temporary correction.

Yellow flame tips. Too little primary air is the problem in this instance. The primary air opening may be either too small or clogged with dirt. Check to see that the proper type and size of inlet fitting is installed. With LP gas, small yellow tips are not necessarily a problem.

THERMOCOUPLES

As stated previously, if the pilot flame is not lit or if there is no gas supply to the pilot burner, the main gas valve must remain closed. If not closed, gas collects in the furnace area, creating a dangerous, explosive atmosphere. A thermocouple is the safety device used to prevent this dangerous situation, Figure 5-7.

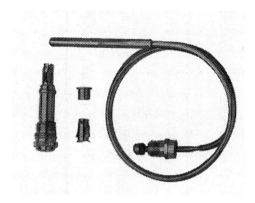

Figure 5-7. Typical Thermocouple. Courtesy , White-Rodgers Division, Emerson Electric Company.

A thermocouple is a bond of two different metals which, when heated, generates a small dc voltage. The voltage produced is measured in millivolts, or 1/1,000 V (0.001 V). While some regular volt-ohmmeters are capable of measuring this dc voltage, special millivoltmeters are available to specifically measure this voltage. Like the volt-ohmmeter shown in Chapter 1, these meters have two leads. They are used and read in exactly the same manner.

HOT AND COLD JUNCTIONS

A thermocouple has a tubular piece surrounding an inner solid element. These two pieces are welded together at one end. This area is heated by the pilot burner flame and called the hot junction. The outer element is brazed to a brass connector sleeve, which in turn, is brazed to a copper tube. Inside the copper tube is an asbestos-insulated copper wire welded to the inner element of the thermocouple. The outer tube and inner wire form the cold junction of the thermocouple. When the hot junction is heated by the pilot and the cold junction remains at a lower temperature, an electrical current is created. The amount of current produced is, within limits, proportional to the temperature difference between the hot and cold junctions. The current is fed to an electromagnet whose magnetic field opens the pilot gas valve and allows gas to flow. As long as heat is applied to the hot junction, current flows and the pilot valve remains open, Figure 5-8.

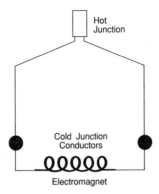

Figure 5-8. Schematic Showing Thermocouple Operation

Since the hot junction must heat up before it can generate enough voltage to hold the pilot gas valve open, a plunger is provided to manually open the valve and allow gas to flow to the pilot burner. Once the pilot ignites, the plunger must be held in manually for about 60 seconds to allow the generated voltage to build up to the point where the coil will hold the valve open.

The generated voltage drops to zero if the pilot goes out. As a result, the valve automatically closes and does not allow any gas to enter the pilot or burner. If the thermostat were to call for heat at this time, the main gas valve would open, but no gas would be available for the main burners. To restore service in this situation, the pilot must be manually reset.

SAFETY SHUTOFF

An added safety factor with natural gas, and a requirement for LP gas, is this system provides 100 percent safety shutoff. A safety shutoff is most important with liquid propane, because LPG is heavier than air, and if unburned gas were delivered to the furnace, it would drop to the floor and present a highly dangerous explosion hazard. Natural gas, being lighter than air, tends to rise and mix somewhat with room air. An important feature of this safety circuit is that it is completely independent of any outside power source. This means that as long as the pilot flame stays lit, the safety circuit functions normally, even when all power is cut off from the furnace.

CHECKING THE THERMOCOUPLE

The thermocouple should be checked under closed circuit conditions. For this check, a General Controls Adapter should be used to provide connection points for the millivoltmeter, allowing the circuit millivoltage to be read under load. The thermocouple can be checked by following these steps:

1. Remove the thermocouple from the gas valve and screw the adapter, fingertight, into the hole.
2. Attach the thermocouple to the top of the adapter, fingertight plus one-quarter turn with a small wrench.
3. Place the positive lead of the multitester on the outside conductor of the thermocouple and the negative lead to the tab of the adapter, Figure 5-9.
4. Use the 0-50 millivolt scale.
5. Reverse the probes if the meter moves to the left of zero or no reading is indicated.
6. Make sure the pilot is burning when readings are taken.

Figure 5-9. Connecting Millivoltmeter Probes to Adapter

When the reading is less than 7 millivolts, perform the following steps:

1. Adjust the pilot gas for a larger flame. At the gas valve, turn the pilot flame adjustment screw counter-clockwise.
2. Clean the primary air holes in the pilot burner.
3. Clean the pilot burner orifice.
4. Replace the thermocouple if the reading is still less than 7 millivolts.

GAS VALVES

Pressure regulators, pilot valves, and main gas valves, Figure 5-10, are a few of the most commonly used gas valves. Each valve has a tap for taking pressure readings and a choice of inlet and outlet connections. They also have a screw for adjusting manifold pressure and pilot flame size. A threaded connection attaches the thermocouple to the gas valve; spade or screw terminals are provided for the main gas valve coil connections.

An arrow on the valve body indicates the direction of flow through the gas valve. Several outlet options are often available, some of which include a left, right or straight-through outlet to the manifold. Good quality pipe dope should be used on the male connection, leaving the last two threads bare.

Figure 5-10. Typical Gas Valve. Courtesy Carrier Corporation, a Subsidiary of United Technologies Corporation.

PILOT BURNER GAS SUPPLY TUBE

The pilot burner gas supply tube is connected with a compression fitting. To connect this compression fitting:

1. Insert the tube into the tapped hole until it reaches the bottom.
2. While holding tubing all the way in, slide the compression fitting into place and engage the threads, turning until fingertight. Using a wrench, tighten one more turn.
3. After installation, check all pipe joints, pilot gas tube connections and valve gaskets for leakage.
4. Brush pipe joints with soap and water solution, and, while the main burner is in operation, watch for bubbles. If there is any sign of bubbles, repair leaks immediately.

CAUTION: Never jumper coil terminals on electrical connections as this shorts out the valve coil and can burn out the heat anticipator in the thermostat.

NOTE: Pilot tubing must always be aluminum and never copper. Natural copper reacts with copper and causes 1/4-inch copper tubing to become completely plugged in a short period of time.

TWO-STAGE HEATING THERMOSTATS

Some valves provide for a two-stage heating thermostat. The first stage activates a preset, low fire position. When the second stage calls for heat, the valve opens to the high fire position. A heat motor in the valve operator accomplishes the shift. This type of valve requires about 30 seconds to recycle.

The simplified block diagrams show the basic gas valve sequence of operation as follows:

1. When the dial is in off position, Figure 5-11, no gas can flow to either the pilot or main burner section.

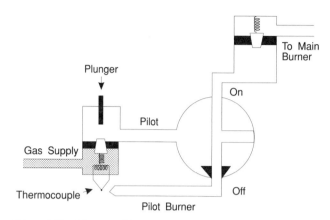

Figure 5-11. Schematic of Gas Valve in Off Position

2. In the pilot position, Figure 5-12, gas can flow through the dial passageway to the pilot burner valves. Depressing the plunger opens the valve manually, allowing gas to flow to the pilot burner. However, even after a call for heat opens the main gas valve, the main burners cannot come on, because use no gas can flow to the main gas valve section. After about 60 seconds, the coil is energized and holds the pilot gas valve open.

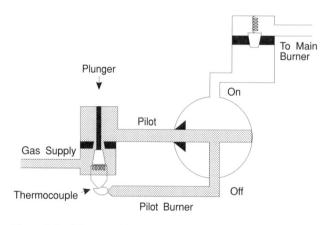

Figure 5-12. Schematic of Gas Valve in Pilot Position

3. When in the on position, Figure 5-13, gas flows through the dial passageway to the pilot burner valve (which is open when the pilot is lit) and to the main burner valve. When the thermostat calls for heat, the main burner valve opens, supplying gas to the main burners for ignition by the pilot.

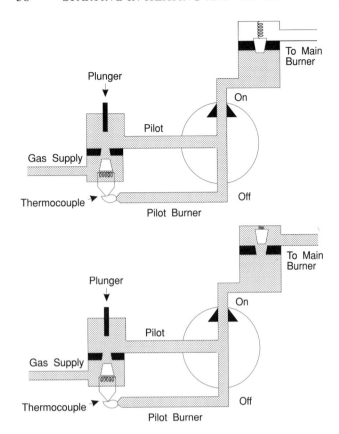

Figure 5-13. Schematic of Gas Valve Before Calling for Heat (above) and After a Call for Heat (below)

SAFETY DROPOUT

A stop watch, or wrist watch equipped with a second hand, is required to check whether or not the safety dropout in the gas valve is functioning correctly. Timing begins when the pilot is turned to the off position. A click indicates when the gas valve closes off the main gas supply. The maximum allowable safety dropout time for a 400,000 Btu gas input furnace is 3 minutes. If the dropout time exceeds 3 minutes, the valve is sticky and should be replaced.

A millivoltage meter should be attached to the thermocouple adapter during this check. A reading below 4 millivolts is normal. If the millivoltage reading is not 4 millivolts or below when safety dropout occurs, the millivoltage coil in the main gas valve is becoming weak. In most cases, the valve must be replaced to correct this problem.

After making the check, the meter leads are removed and the thermocouple adapter removed from the gas valve and thermocouple. The thermocouple in the gas valve must be replaced (tightened fingertight plus one-quarter turn with a light wrench). The pilot can now be lit and the gas valve turned to the on position.

MANIFOLD GAS PRESSURE

The combination gas valve contains a pressure regulator. This is to ensure the burners receive constant gas pressure, even when the mains vary. While the gas valve opens relatively quickly, the regulator opens slowly. This keeps the pilot from starving and the light-up quiet.

Every service technician's tool kit requires a simple manifold pressure test kit. It should contain a B valve or petcock which screws into the test port on the gas valve or manifold, a 4-foot length of rubber tubing, and a pressure gage.

Manifold gas pressures are measured in inches of water column (W.C.), as these pressures are less than 1 lb/square inch (psi). (One inch of water column pressure is the pressure needed to push a column of water, usually in a manometer, up one inch.) For natural gas, manifold pressure should be between 3-1/2 and 4 inches W.C.; for propane, 11 inches W.C.

To check manifold pressure, perform the following steps:

1. Remove all power from the gas valve by turning off the furnace disconnect switch. This prevents the gas valve from opening and allowing gas to leak into the room while the plug is out.
2. Remove the pipe plug that is located on the gas valve or manifold. With the valve in the off position , install the B valve into this plug tap.
3. With the hose fitting attached to the B valve, take a pressure reading with either a U-tube manometer or manifold gas pressure gage, Figure 5-14.

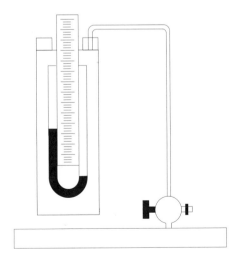

Figure 5-14. U-Tube Manometer Attached to B Valve

MANOMETERS

The U-tube manometer has a graduated scale in the center of the U and shutoff tubing connectors which open into the top of each leg. These can be opened by twisting one turn counterclockwise. The manometer has a magnet on the assembly so it can stick to any smooth steel surface, or it can be hung vertically. When measuring gas pressure, the U-tube contains water, the height of which is the same in both tubes when equal pressure exists. The center scale is adjusted so that the zero mark is set at this point.

The rubber or nylon tubing connects one of the shutoff openings to the B valve. When the valve is opened, gas pressure is applied to one column water, forcing that column down and the other column up. This latter column is open to the atmosphere. The difference in pressures—the sum of the amount one column is below zero and the amount the other is above—is the pressure reading, Figure 5-15. It is always important to turn the gas off when connecting the manometer.

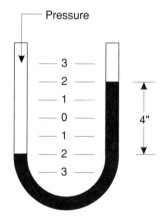

Pressure

3
2
1
0
1
2
3

4"

Figure 5-15. Manometer Reading

On some manometers, the zero setting is located at the bottom of the scale with numbers only on one side. With this type of manometer, the B valve connects to the opening on the side that is not numbered. The pressure, in inches W.C., is read directly from the height of the numbered column.

PRESSURE REGULATORS

In order to check the pressure reading, it is necessary to turn on the disconnect switch and open the B valve when the thermostat calls for heat. As stated previously, the natural gas pressure reading should be between 3 and 4 inches W.C.; LPG, 11 inches W.C. The pressure regulator needs adjusting if the readings are above or below these points. Gas valves with a built-in pressure regulator have an adjustment screw on the gas valve to change the pressure. This screw can be

turned up or down to increase or decrease pressure; one-quarter turn or less is normally sufficient to correct the pressure reading. A unit with a separate pressure regulator has an adjustment screw on the top which can be turned in the same manner.

Unlike natural and other gases, LPG does not normally use a pressure regulator. This is due to the LPG supply tank's requirement that a master regulator be present in order to maintain a constant downstream pressure of 10 to 12 inches W.C. As this regulator is always installed, it is not necessary to duplicate its function at the furnace.

After the reading is taken and the pressure regulated, the disconnect should be turned off, the gage, B valve and U-tube removed, and the plug put back in the gas manifold. It is then necessary to apply sealing compound to the plug. A double-check should be made at this point to ensure no leaks exist.

CONTROL CIRCUIT AMP-DRAW

As described in Chapter 2, an amperage multiplier can be used to check the control circuit amp-draw at the thermostat. The same value can be taken at the gas valve (on upflow furnaces only). This value can be obtained by performing the following steps:

1. Remove one of the low voltage wires from the gas valve terminal.
2. Connect one end of the amperage multiplier to the removed low voltage wire and the other end to the terminal from which the wire was disconnected.
3. Turn the furnace disconnect on and record the reading with an Amprobe by opening its jaws and placing them around the coil of the amperage multiplier.
4. Divide this reading by ten in order to obtain the actual amperage.
5. Turn the disconnect switch off, remove the amperage multiplier, and reconnect the low voltage wire to the gas valve.
6. Turn the disconnect switch on.

TEMPERATURE RISE

The number of degrees the air is heated as it passes through the furnace is known as the temperature rise. Most gas furnaces allow a temperature rise of 80 to 100 °F. Inserting a thermometer in the return air side of the furnace and then reading its temperature gives an accurate temperature rise measurement. Normally, this measurement is in between 68 to 72 °F.

A second reading is taken in the supply air plenum. The supply air reading should be taken at a point that is not in the line-of-sight of the heat exchanger. This is to ensure that the radiant heat from the heat exchanger does not influence the supply air reading. The supply air temperature can also be taken at a supply air register. To compensate for heat loss in the ducts, it is necessary to allow 1/2 °F/ft of duct from the plenum.

The supply air is normally about 90 °F warmer than the return air. If the temperature rise is below 90 °F, then the blower speed should be decreased in order to increase the heat rise through the furnace. If the rise is greater than 90 °F, then the blower speed should be increased to reduce the rise. One full turn of the outer half of the adjustable pulley changes the temperature rise through the furnace about 10 to 15 °F.

GAS PIPING

Piping to the gas unit must be of the correct size. The pipe should run directly from the meter to the unit with as few bends as possible. The riser from the burner should consist of a tee fitting with the bottom outlet capped and extended to form a drip leg. This prevents dirt and dust from being carried into the burners , Figure 5-16. During installation, all piping should be cleaned of dirt and scale. Carefully reaming the ends of piping and tubing is necessary to remove obstructions and burrs.

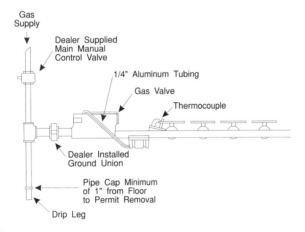

Figure 5-16. *Gas Piping Layout Including Drip Leg*

Nonhardening pipe joint compound should only be used sparingly on the male threads. As these joints must be gas tight, using a soap solution helps to check for leaks.

A manometer can also be used to help check for leaks, and the following steps outline this procedure:

1. Place the manometer in the system as usual.
2. Turn on the gas and read the pressure in inches W.C.
3. Turn the gas off. The pressure should remain the same as when the gas was turned on. If it does not, this indicates a leak. It is a good idea to leave the manometer in this position for several hours, just to make sure there is no leak.
4. If a leak is indicated, use the soap solution to help find it.

If there are no leaks, the system must be purged. This means the gas must be bled off in order to get rid of air and other gases within the system. The service technician must use extreme caution when performing this procedure, as gas is explosive. Also, the purging should take place in a well-ventilated area, so no gas can collect in the area of combustion.

PIPE SIZES FOR GAS HEATERS

When specifying the proper size of gas piping for gas heaters, the recommendations listed in Table 5-1 are normally used. (It is always important for the service technician to first check all local codes before proceeding.) These recommendations are based on piping designed for a pressure drop, from the meter to the heaters, of 0.3 inches W.C. The capacities, in ft^3/hour, are shown in Table 5-1 for various pipe sizes and lengths. These capacities are based on using gas with a specific gravity of 0.60. (If the gas being used has a different specific gravity, adjust the value from Table 5-1 by the multiplier in Table 5-2.)

Length of Pipe in Feet	Normal Diameter of Pipe (inches)				
	1/2	3/4	1	1-1/4	1-1/2
15	76	172	345	750	1220
30	55	120	241	535	850
45	44	99	199	435	700
60	38	86	173	380	610
75		77	155	345	545
90		70	141	310	490
105		65	131	285	450

Table 5-1. *Pipe Capacity Cubic Feet of Gas per Hour*

Specific Gravity	Multiplier
.35	1.31
.40	1.23
.45	1.16
.50	1.10
.55	1.04
.60	1.00
.65	0.962
.70	0.926
.75	0.895
.80	0.867
.85	0.841
.90	0.817
1.50	0.633
1.55	0.622
2.00	0.547

Table 5-2. *Multipliers for Table 5-1 When the Specific Gravity of Gas Is Other Than 0.60*

Adopting the conservative pressure drop of 0.3 inches W.C. allows for the extra resistance found when the piping uses the ordinary number of fittings. To determine the length and carrying capacity of the gas piping needed from the information given in Table 5-1, it is only necessary to measure the length of the gas piping from the meter to the heater. It is not necessary to account for the extra resistance imposed by the normal number of fittings.

A pressure drop greater than 0.3 inches W.C. is permissible in areas where high gas pressure is always available at the house service. However, the pressure drop through the heater controls must never exceed 1 inch W.C. If 8 inches W.C. of pressure is always available at the meter, piping designed for as much as 1 inch W.C. pressure drop would still have adequate pressure available at the heater.

LP GAS TANKS AND PIPING

Butane and propane, both LP gases, have specific gravities of 1.53 and 2.01, respectively. Both are shipped and stored in liquid form under approximately 200 lbs of pressure. Before they can be used in a furnace, they

must be vaporized (this process is described below). As stated previously, an LP installation does not require a pressure regulator. In their Manual 58 safety standards, the National Fire Protection Association covers LP gas equipment installation.

Using Table 5-1 and applying the multiplier from Table 5-2 (to account for the different specific gravity for LP gas) can determine the piping size needed. Tables 5-3 and 5-4 offer an alternate method of calculating the piping sizes.

Use this size cooper tubing or standard pipe to keep pressure drop below 2 lbs for the maximum flow shown if the line between regulators (tank to building) is the long run.

Capacity	25 Feet	50 Feet	75 Feet	100 Feet
50 cfh 125,000 Btuh	3/8" od Tubing	3/8" od Tubing	3/8" od Tubing	3/8" od Tubing
100 cfh 250,000 Btuh	3/8" od Tubing	3/8" od Tubing	3/8" od Tubing	1/2" od Tubing
150 cfh 375,000 Btuh	1/2" od Tubing	1/2" od Tubing	1/2" od Tubing	1/2" od Tubing
200 cfh 500,000 Btuh	1/2" od Tubing	1/2" od Tubing	1/2" od Tubing	1/2" od Tubing
300 cfh 750,000 Btuh	1/2" od Tubing	3/4" Pipe	3/4" Pipe	3/4" Pipe

Table 5-3. *Alternate Method No. 1 of Calculating Piping Size*

Use this size cooper tubing or standard pipe to keep pressure drop below 2 lbs for the maximum flow shown if the line between regulators (tank to building) is the long run.

Capacity	10 Feet	20 Feet	30 Feet	40 Feet	50 Feet
10 cfh 25,000 Btuh	3/8" od Tubing	3/8" od Tubing	1/2" od Tubing	1/2" od Tubing	1/2" od Tubing
20 cfh 50,000 Btuh	1/2" od Tubing	1/2" od Tubing	1/2" od Tubing	5/8" od Tubing	5/8" od Tubing
30 cfh 75,000 Btuh	1/2" od Tubing	5/8" od Tubing	5/8" od Tubing	5/8" od Tubing	5/8" od Tubing
50 cfh 125,000 Btuh	5/8" od Tubing	5/8" od Tubing	3/4" Pipe	3/4" Pipe	3/4" Pipe
75 cfh 187,500 Btuh	3/4" Pipe	3/4" Pipe	3/4" Pipe	3/4" Pipe	3/4" Pipe

Table 5-4. *Alternate Method No. 2 of Calculating Piping Size*

PROPER GAS PRESSURE

Maintaining proper gas pressure depends upon three main factors: (1) vaporization rate, (2) proper pressure regulation, and (3) pressure drop in lines.

Vaporization Rate. Outside air temperature and the amount of fuel in the tank determine the vaporization rate. Heat passes through the walls of the tank via the outside air. This heat comes into contact with the liquid fuel (LPG) in the tank, causing the LPG to vaporize. This heat transfer occurs at a greater rate when there is a high liquid level in the tank. Therefore, high liquid levels are required for adequate vaporization. When the tank is nearly empty, very little heat transfer occurs, and vaporization becomes inadequate.

When air temperature around the tank becomes cold, the size of the tank plays an important role. If the tank is too small, there is not enough tank wall area to collect the heat required to vaporize all the fuel. Table 5-5 shows the sizes of tanks required to vaporize the gas consumed at various winter temperatures. These ratings assume that the tank is at least half full. Referring to Table 5-5, notice that a 250 gallon tank vaporizes 100 ft^3 of gas/hour at 10 °F and that this drops off to 50 ft^3 of gas/hour at -10 °F.

Capacity	32 °F	20 °F	10 °F	0 °F	-10 °F	-20 °F	-30 °F
50 cfh 125, 000 Btuh	115 gal	115 gal	115 gal	250 gal	250 gal	400 gal	600 gal
100 cfh 250,000 Btuh	250 gal	250 gal	250 gal	400 gal	500 gal	1000 gal	1500 gal
150 cfh 375,000 Btuh	300 gal	400 gal	500 gal	500 gal	1000 gal	1500 gal	2500 gal
200 cfh 500,000 Btuh	400 gal	500 gal	750 gal	1000 gal	1200 gal	2000 gal	3500 gal
300 cfh 750,000 Btuh	750 gal	1000 gal	1500 gal	2000 gal	2500 gal	4000 gal	5000 gal

Table 5-5. Recommended tank size for the lowest outdoor temperature (average for 24 hour period)

Proper Pressure Regulation. Many LPG installations use the single-stage pressure regulation system. The single-stage system uses one regulator at the tank to reduce tank pressure from about 200 psig to the 11 inches W.C. required at the furnace, Figure 5-17. This system creates problems, because a small vibration occurring between the tank and house can cause the pressure to drop below 11 inches W.C. This upsets the normal operation of the furnace. In addition, moisture in the fuel can freeze due to the small outlet required.

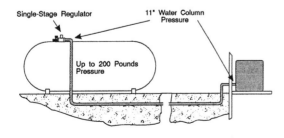

Figure 5-17. LP Gas Installation with Single-Stage Regulator

The two-stage pressure regulation system, Figure 5-18, consisting of a first and second regulator, is a better method. The first regulator, at the tank, reduces the 200 psig in the tank to between 5 and 15 psig in the intermediate line. This is important, because Manual 58 of the NFPA states the maximum pressure that can be taken into a building is 20 psig. An intermediate line pressure of about 10 psig provides sufficient capacity. The first regulator must have strong parts (i.e., a strong diaphragm and a large orifice built for handling lbs-to-lbs pressure) and be capable of having up to 250 psig inlet pressure and 20 psig outlet pressure.

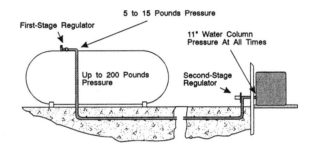

Figure 5-18. LP Gas Installation with Two-Stage Regulator

The second regulator should have a 3/16 or 1/4 inch orifice with an inlet pressure of approximately 10 psig and an outlet pressure of 11 inches W.C. It must be 5 feet away from the closest lower opening to the house, and the vent should be turned downward, Figure 5-19.

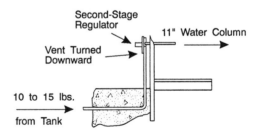

Figure 5-19. Installing Second Regulator

The first and second regulators in a two-stage system work together to give more uniform operating pressure, reduce the possibility of freezing and reduce the size of the piping from the tank to the house. Therefore, the two-stage system is preferable to the one-stage system.

Pressure Drop in Lines. Finally, it is important to consider the pressure drop in the lines between the first and second regulators, and also between the second regulator and the furnace. Taking these pressure drops into consideration helps determine the proper pressure necessary.

LPG PIPE DOPE

Because LPG is an excellent solvent, special pipe dope must be used when assembling an LPG piping system. LPG dissolves white lead and most standard commercial compounds, so shellac-based compounds are recommended for use. During peak periods, many gas companies mix LPG with other gases, so it is good practice to use the aforementioned compounds for any type of gas piping.

BURNER INPUT

The flow of gas at the gas meter can be timed and converted to Btu input/hour in order to measure the input to a gas heater. When measuring, all other gas appliances in the house must be turned off to ensure an accurate reading. There is one pointer on a gas meter that measures small quantities of flow such as 1, 2 or 5 ft^3, Figure 5-20. For each complete revolution of the pointer on that dial, the marking on the dial tells how many ft^3 of gas pass through the meter.

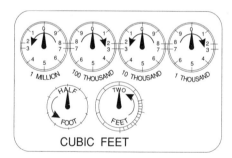

Figure 5-20. Typical Gas Meter Dials

To obtain Btu input:

1. Turn the heater on.
2. Determine how much time is required for one revolution on the smallest dial (Table 5-6 shows the ft^3/hour the furnace uses).
3. Multiply this time by the calorific value of the gas. (Remember that heater output is usually taken as 80 percent of the input.)

The local gas company can supply information regarding the average Btu/ft^3 of the gas in the installation area. Using Table 5-6, if the smallest dial measures 1 ft^3, and with only the furnace on, it takes 38 seconds for the pointer to make one revolution, then the furnace consumes 95 ft^3 of gas/hour. If the heat content of this gas is 1000 Btu/ft^3, then the furnace input is 95,000 Btu, and the adjusted output is 76,000 Btu (0.8 x 95,000).

FIELD WIRING

When field wiring, the first step is to run a 120-V, two-wire service from the main breaker panel in the house to the furnace. The main circuit breaker or fuse should be sized in accordance with the instructions that come with the unit. Wire size is specified in the instructions and must be in accordance with the National Electric Code (NEC). Some areas require conduit for all or part of this service, so local codes must always be checked.

A disconnect switch (this can be a fused handy box) with an off-on switch must be provided on or near the furnace. This is fused according to manufacturer's instructions. OR OUTLET PLUG

To perform field wiring for a heating-only gas furnace:

1. Connect the two wires to the line side of the box with the fuse in the hot leg (black wire).
2. Run two wires from the disconnect to the line voltage make-up box in the furnace.
3. Connect the black wire to the fan control and transformer leads.
4. Connect the white wire to the white leads from the blower and transformer, Figure 5-21.

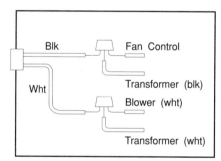

Figure 5-21. Make-up Box Wiring Connections

5. Run two low voltage wires (three if there is fan control at the thermostat) from the thermostat to the low voltage make-up box or terminal block in the furnace. The terminal block is marked in the same manner as the thermostat. Thus R, W and G (if used) from the thermostat connect to the R, W and G terminals of the terminal block.

Seconds For One Revolution	Size of Test Meter Dial (ft³)					Seconds For One Revolution	Size of Test Meter Dial (ft³)			
	One-half	One	Two	Five			One-half	Qne	Two	Five
10	180	360	720	1,800		50	36	72	144	360
11	164	327	655	1,636		51	35	71	141	353
12	150	300	600	1,500		52	35	69	138	346
13	138	277	555	1,385		53	34	68	136	340
14	129	257	514	1,286		54	33	67	133	333
15	120	240	480	1,200		55	33	65	131	327
16	112	225	450	1,125		56	32	64	129	321
17	106	212	424	1,059		57	32	63	126	316
18	100	200	400	1,000		58	31	62	124	310
19	95	189	379	947		59	30	61	122	305
20	90	180	360	900		60	30	60	120	300
21	86	171	343	857		62	29	58	116	290
22	82	164	327	818		64	29	56	112	281
23	78	157	313	783		66	29	54	109	273
24	75	150	300	750		68	28	53	106	265
25	72	144	288	720		70	26	51	103	257
26	69	138	277	692		72	25	50	100	250
27	67	133	267	667		74	24	48	97	243
28	64	129	257	643		76	24	47	95	237
29	62	124	248	621		78	23	46	92	231
30	60	120	240	600		80	22	45	90	225
31	58	116	232	581		82	22	44	88	220
32	56	113	225	563		84	21	43	86	214
33	55	109	218	545		86	21	42	84	209
34	53	106	212	529		88	20	41	82	205
35	51	103	206	514		90	20	40	80	200
36	50	100	200	500		94	19	38	76	192
37	49	97	195	486		98	18	37	74	184
38	47	95	189	474		100	18	36	72	180
39	46	92	185	462		104	17	35	69	173
40	45	90	180	450		108	17	35	67	167
41	44	88	176	440		112	16	32	64	161
42	43	86	172	430		116	15	31	62	155
43	42	84	167	420		120	15	30	60	150
44	41	82	164	410		130	14	28	55	138
45	40	80	160	400		140	13	26	51	129
46	39	78	157	391		150	12	24	48	120
47	38	77	153	383		160	11	22	45	112
48	37	75	150	375		170	11	21	42	106
49	37	73	147	367		180	10	20	40	100

Table 5-6. *Gas Input to Burner in Cubic Feet per Hour*

To convert to Btu per hour, multiply by the Btu heating value of the gas used.

REVIEW QUESTIONS

1. Describe the functions of a heat exchanger. DISTRIBUTE HEAT, FLUE
2. What is a simple test for heat exchanger leakage? SALTWATER SPRAY
3. How does a burner receive primary and secondary air? SHUTTERS
4. In what way is the amount of gas controlled in the burner? REGULATOR + ORIFICE
5. What is the function of the burner venturi? PULL PRIMARY AIR
6. How can several burners be lighted at the same time? CROSSOVER
7. What ignites the main burners? PILOT
8. What is the function of a thermocouple? PILOT SAFETY
9. Give an example of properly and improperly operating flames. PROPER – BLUE
10. What is a millivoltmeter?
11. If the power is lost to the furnace, what happens to the safety circuit? STAYS IN EFFECT
12. How is a low millivoltage situation corrected? CLEAN PILOT, ADJUST TO LARGER FLAME
13. Give the sequence of the combination gas valve components.
14. List the steps for checking a good piping installation. MANOMETER
15. What is meant by a safety dropout? TIME IT TAKES FOR GAS VALVE TO SHUT OFF
16. How is manifold pressure checked? MANOMETER IN GAS VALVE PORT
17. Describe a U-tube manometer.
18. What is considered to be a normal temperature rise for a gas furnace? 85° TO 100°
19. What are the three factors affecting gas pressure for an LPG system?
20. Describe the advantages of a two-stage regulator for an LP system.

Chapter **6**
Oil Heating

An oil furnace has some special requirements which influence the design of the heat exchanger and the oil burner. For example, in order for combustion to take place, an oil burner must atomize the oil and mix it with the proper amount of air. In addition, an oil burner must maintain a fairly high temperature around the flame at all times, or incomplete combustion results. Oil furnace designs must consider these factors and provide a means of dealing with them. A typical oil furnace is shown in Figure 6-1.

Figure 6-1. *Typical Oil Furnace. Courtesy, Bard Manufacturing Company.*

OIL CHARACTERISTICS

As described in Chapter 3, oils are classified according to viscosity (thickness). The thicker an oil, the more resistant it is to flow and subsequently the higher its viscosity. This viscosity determines its grade (on a scale of 1 to 6 with 6 being the thickest).

When an oil has a higher viscosity, it contains more Btu/gallon. For example, No. 2 oil contains 144,000 Btu/gallon, while No. 6 oil contains 152,000/gallon. Most domestic oil burners use No. 2 oil, because this grade of oil is less expensive and has good lubricating properties. However, if the oil tank is located outside in extremely cold weather for a long period of time, then No. 1 oil is recommended. This is due to the fact that cold weather causes the oil to become thick and improperly atomized. Hence, the burner, when it receives cold oil, produces long, narrow, noisy flames that burn on the back wall of the combustion chamber rather than directly in front of the burner. This results in impingement. Gasoline and crankcase oil are not recommended for a domestic oil burner.

OIL STORAGE

If the oil storage tank is buried outside, all the pipe connections must be made with swing joints so that no breakage occurs if the tank settles. The top of the tank should be below the frost line. The tank vent should slant toward the tank and be at least 1-1/4 inch pipe size. The vent should terminate outside the building, at a point 2 or more feet (vertically or horizontally) away from any window or other building opening. The vent should rise above the normal accumulation of snow and ice and have a weatherproof hood. Interior tanks should not be located within 7 feet (horizontally) of any fire or flame.

All pipe work and fittings must be airtight and only high grade materials used. Pipe lines should be run as directly as possible, free of traps, and placed out of the way; if possible, beneath the floor. Copper tubing is the most desirable material for this application. For relatively short runs on gravity-flow inside tanks, 3/8 inch outside diameter (od) is the most common; longer runs require 1/2 inch od. On outside, underground tanks where there is a suction lift, 1/2 inch od tubing is recommended.

On all installations requiring suction lift, the return line must run at the proper angle to ensure entrained air, in the oil or in the system, is returned to the tank. This is necessary for proper fuel pump operation. The fill pipe should measure 2 or more inches and slant toward the tank. Termination of the fill pipe outside the building at a point not less than 4 feet from any building opening at the same or a lower level is desired. Fuel terminals should be closed tight, provided with a metal cover, and designed to prevent tampering. The tank has a shutoff valve at the exit and a filter between the shutoff valve and the furnace.

The tank may either have an oil level indicator or a plug, allowing the oil level to be checked with a dipstick. A standard size vertical tank holds 275 gallons of fuel oil. Inserting a dipstick from the top shows the approximate number of gallons in the tank, Figure 6-2. When determining the amount of water in the tank, it is helpful to place gray litmus paste on the bottom of the dipstick. As litmus paste turns purple in the presence of water, it is easy to determine the water level by measuring this area. If the water level exceeds 1 inch, it should be drained out of the tank.

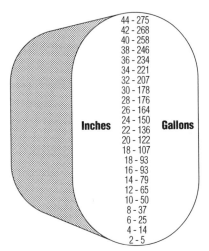

Figure 6-2. *Oil Level Table Equates Gallons to Inches on Dipstick*

When the oil tank is above the burner, or when the lift does not exceed 8 feet, a single-pipe gravity system is used. In a single-pipe system, one line of 3/8-inch copper tubing runs from the oil tank shutoff valve to the oil burner, Figure 6-3. Excess oil circulates internally in the pump, back to the suction side. In this gravity system, the allowable length of 3/8-inch od tubing is 100 feet; for a 5- foot lift, the allowable length is 90 feet; with an 8-

foot lift, this drops to 70 feet. Any system may use a two-pipe system; however, when the vertical lift exceeds 8 feet, the two-pipe system must always be used , Figure 6-4.

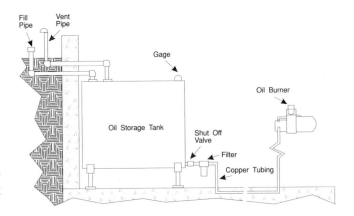

Figure 6-3. *Single-Pipe Gravity Feed System with Oil Tank Above Burner*

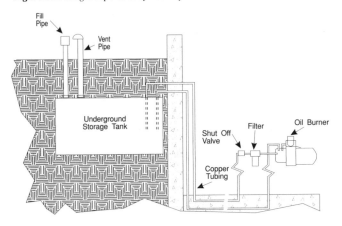

Figure 6-4. *Two-Pipe System Used When Vertical Lift Exceeds 8 Feet*

The tank receives the return line. If the bottom of the tank is below the pump intake, the return line should be inserted three to four inches from the tank bottom. If the bottom of the tank is higher than the pump intake, the return line should not extend more than 8 inches inside the tank. Allowable line lengths for various amounts of lift are shown in Table 6-1.

One-Line Lift / One-Stage Pump		Two-Line / Two-Stage Pump	
Lift (ft)	Length (ft)	Lift (ft)	Length (ft)
5	90	0	65
8	70	1	60
		2	54
		3	50
		4	45
		5	40
		6	35
		7	30
		8	25
		9	20
		10	16

Table 6-1. *Allowable Line Lengths for Different Lifts*

OIL FILTERS

All installations have an oil line filter located between the shutoff valve from the tank and the furnace. Most filters have a cartridge which can (and should) be replaced periodically to make sure that dirt and other foreign materials do not enter the furnace burner. To replace a filter cartridge, the following steps should be performed:

1. Turn off the hand valve to shut off the oil from the tank, and place empty pan under filter to catch spill-over.
2. Unscrew the bolt on the top of the filter assembly, remove the bowl, then remove the cartridge from the bowl.
3. Clean the inside of the bowl and remove the gasket (a new gasket is always supplied with a new cartridge).
4. Place the new cartridge into the bowl, set the new gasket on the lip and tighten the mounting bolts firmly.
5. Open the hand valve when replacement is complete.
6. Most filters include a bleed port to allow air entering the cartridge during the change to bleed off, thus preventing air from entering the system. When opening the oil valve, slowly open the bleed port at the same time. This enables air to bleed out of the filter. As soon as oil appears at the bleed hole, close the bleed port.
7. Wipe up any spill-over.

HEAT EXCHANGERS

Heat exchangers have a primary heating surface, or combustion chamber, and a secondary heating surface. This configuration is necessary in oil heating, because the combustion chamber must come up to temperature rather quickly to support combustion. Air blows across this secondary area, which is a metal surface, and absorbs the heat from the flame.

Most combustion chambers are round, square or rectangular. As they produce the least amount of air turbulence within the chamber itself, round chambers are possibly the most popular design. Round chambers can be positioned either vertically or horizontally. Several shapes of oil heat exchangers are shown in Figure 6-5.

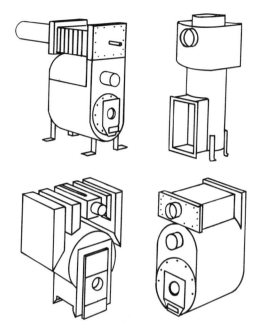

Figure 6-5. *Various Types of Oil Heat Exchangers*

Either a light firebrick, or special wraparound, blanket-type insulating material line the combustion chamber. The latter material has several trade names. This material is lightweight, quite thin and can withstand temperatures up to 2,500 °F, which is well above the operating range of most domestic oil furnaces. This material takes only about 10 seconds to come up to temperature, and it absorbs very little heat. To use a refractory in the combustion chamber ensures the temperature in the chamber remains high. This helps in complete combustion of the oil and prevents impingement on a cold surface.

The secondary heating surface can either be round or square. It contains a series of baffles and channels so that the heated air and products of combustion follow a twisting path, thereby heating the entire surface evenly. To ensure efficient heat exchange, the secondary heating surface is also designed to allow contact of the supply air with the entire outer surface.

A furnace is equipped with an inspection hole, which allows the service technician to observe the size, shape, and direction of the flame without opening the door. This peephole has a cover which can be slid aside when it is necessary to observe the flame. The door is also hinged, allowing the service technician, by way of an inspection mirror, to observe the shape of the flame as it leaves the burner nozzle.

Oil heat exchangers are made from cold-rolled or aluminized steel. In many cases, they are coated with a ceramic to provide longer life in corrosive atmospheres. Most heat exchangers have access doors so that the heat exchanger itself can be cleaned of any soot or carbon residue. Both the primary and secondary heat exchangers have cleanout doors.

FAN AND LIMIT CONTROLS

Oil furnaces contain either individual fan and limit controls, or a combination fan and limit. The information contained in the gas heating chapter (Chapter 5) concerning fan and limit controls also applies to oil heating. One difference between oil and gas is that often both the fan and limit are in the line voltage circuit of an oil furnace.

In an oil furnace, fan controls are set to come on at about 100 °F with a differential of about 25 °F, with the off point being 75 °F. The limit is usually set for 200 °F.

PRIMARY CONTROLS

A primary control contains the safety control and flame detection circuits for an oil burner. The primary control has its own transformer and carries both line and low voltage. An example of a primary control is shown in Figure 6-6. Some primary controls are solid state while others include a solid-state safety device and a relay for blower control.

Figure 6-6. Primary Control. Courtesy, White-Rodgers Division, Emerson Electric Company.

The line voltage side of the primary control is connected to the ignition transformer and burner motor (orange wire). The hot, or black, wire is connected to the limit control (or both limits in the case of a downflow furnace), and the white is connected to neutral. On the low voltage side, there are two terminals marked T and T, or T1 and T, for the thermostat connections (R and W from the thermostat). There are also two terminals for the flame detection device, or cad cell; these terminals are marked S and S, or FD and FD. The low voltage side does not have polarity and therefore, in both cases, either wire can be connected to either terminal. Figure 6-7 is a simplified wiring scheme.

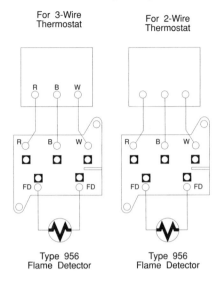

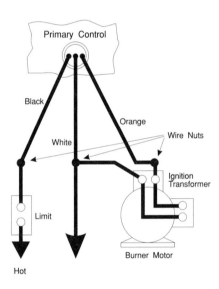

Figure 6-7. Wiring Diagram for the Primary Control. Courtesy, White Rodgers Division, Emerson Electric Company.

During normal operation, limit controls are in the closed position. When the thermostat calls for heat, the motor relay contacts energize the ignition transformer and burner motor, starting the motor and the spark ignition. If ignition results, the flame detector takes the heater, which operates the safety circuit, out of the circuit. The burner then runs until the thermostat is satisfied.

Basic checks on a new installation or a routine call are necessary. These checks are outlined below.

1. Open the disconnect switch at the furnace.
2. Adjust the thermostat above the room setting so that it calls for heat.

3. Make sure there is a supply of oil to the burner.
4. Depress the manual reset button (colored red) on top of the control.
5. Close the disconnect switch and let the burner run for about 5 minutes.
6. Remove one of the flame detector leads from the FD or S terminal. This control should lock out on safety, stopping the burner in approximately 15 seconds. As this time varies, check the manufacturer's controls.
7. Open the disconnect switch.
8. Replace the flame detector lead.
9. Turn on the power (furnace should not start).
10. Wait about 3 minutes, then depress the manual reset button to turn on the burner.
11. Cycle the furnace several times from the thermostat to be sure that it is operating properly.

When a burner starts but then locks out on safety, check the primary control and flame detector as follows:

1. Open the disconnect switch on the line voltage circuit to the furnace.
2. Remove the thermostat wires from the primary control terminals T and T.
3. Place a jumper wire across these terminals to complete the thermostat circuit.
4. Disconnect the two flame detector wires from the primary control.
5. Connect one end of a resistor to one of the flame detector terminals on the primary control. For White-Rodgers and Simicon, this resistor should be 2,000 ohms (red, black, red). For Honeywell it should be 1,500 ohms (brown, green, red). Wattage is not important. Make sure that the other end is shaped in such a way that it can reach the other terminal quickly and easily, Figure 6-8.

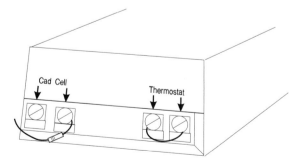

Figure 6-8. *Resistors Wired to Cad Cell Terminals*

6. Close the disconnect switch to the furnace and operate the red manual lever or button to start the burner.
7. Connect the other end of the resistor to the remaining flame detector terminal on the primary control. This must be done quickly to prevent the primary from locking out on safety.
8. If the primary locks out on safety with the resistor in place, consider the primary to be defective. However, be sure that the resistor was connected quickly enough. If any doubt exists, wait 5 to 10 minutes and repeat the test. If the primary does not lock out with the resistor in place, the primary is functioning properly.

FLAME DETECTOR

The oil burner control system must have a way of knowing whether or not ignition is accomplished and a flame established and maintained during the heating cycle. The system must stop the supply of oil and turn off the furnace if a flame is not established. The system senses when this occurs by means of a flame detector, which utilizes a cadmium sulfide cell, or cad cell.

CAD CELL

A cad cell, as stated in Chapter 1, is a light-sensitive device that shuts off the fuel supply if the burner fails to ignite. Cadmium sulfide has the unique capability of being a conductor of electricity in the presence of light in the visible range, yet it resists the passage of electricity in darkness. Thus, the cad cell can determine whether or not a flame is present.

In darkness, the cad cell's resistance to the flow of electrical current in darkness is about 100,000 ohms, which is great enough to prevent any flow. However, in visible light, the resistance drops to less than 1,600 ohms, which allows enough current to flow in order to pull in a sensitive relay in the primary control. The primary then senses ignition and allows oil to continue flowing to the burner. A well-adjusted burner causes the cell to operate in the range of 300 to 1,000 ohms.

The cadmium sulfide cell is located at the rear end of the burner tube in direct alignment with the flame from the nozzle. If the burner ignites, the cad cell sees this flame and starts conducting electricity. The electrical circuit established then takes the heater out of the safety circuit in the primary control, which allows the burner to run until the thermostat is satisfied.

Positioning the cad cell can be tricky, as it must be able to see the light from the flame yet not be 'fooled' by transient light from other sources. The manufacturer carefully positions the cad cell, so its location should not be disturbed.

CHECKING A FLAME DETECTOR

The flame detector can be checked with an ohmmeter set on the Rx10k scale. With the burner off, the flame detector leads are removed from the primary control and the resistance across these leads checked. This resistance should read over 100,000 ohms when the furnace is off. The following steps should be performed when checking a cad cell under normal operation:

1. Jumper thermostat terminals T and T on the primary control.
2. Depress the reset button and close the disconnect switch to start the burner.
3. Disconnect the cad cell leads at the primary control when the burner comes on and is operative. This stops the burner in approximately 45 to 50 seconds. If the burner does not stop, replace the primary control, as it is defective.
4. After 2 or 3 minutes, connect an ohmmeter to the two cad cell leads. Resistance should be 100,000 ohms or better; if not, the cell is defective, or there is light leakage into the burner. Make sure all access openings are sealed against light.
5. Turn off disconnect switch and insert a jumper wire into one of the cad cell connections on the primary control. If both connections are jumpered at this point, the burner will not start.
6. Close the disconnect switch. The burner should not start until the reset lever on the primary control has been set.
7. Once the burner starts, complete the jumper connection to the other cad cell connection on the primary control.
8. With the burner running, connect an ohmmeter to the two leads of the cad cell and read the resistance. This resistance should not be greater than 1,500 ohms, and ideally should be between 600 and 700 ohms.
9. A resistance greater than 1,000 ohms indicates that the cad cell may be dirty, cracked or broken; misaligned in the burner; or have loose or broken wires. Check for these conditions.
10. Carefully remove the cad cell and wipe the face with a soft rag and replace. If the situation still persists, replace the flame detector, as it is faulty. When removing the jumper on the cad cell, be careful not to touch the thermostat wire to the flame detector terminal, as this can damage the primary control.

HIGH-PRESSURE, GUN-TYPE OIL BURNERS

As stated previously, oil must be atomized in order to achieve complete combustion. A high-pressure oil burner combines a number of components into one assembly to accomplish this. They are as follows:

- The burner motor to drive the pump and the blower.
- The combustion air blower.
- The ignition system, consisting of electrodes and transformer.
- The nozzle and burner tube.
- The fuel unit.

The industry has been able to standardize the components required to build high-pressure, gun-type burners. As a result, most parts are assembled on a standard housing or casting. A small number of manufacturers make the pumps, nozzles, transformers, motors and other components, so this makes maintenance and installation problems simple. As manufacturer designs normally resemble one another, a service technician needs to understand only one design in order to service other designs. Should these parts fail, they are normally replaced, rather than repaired, at the job site.

BURNER MOTORS

The burner motor can be either a four-pole, 1,750 rpm motor or a two-pole, 3,500 rpm motor. The shaft connects directly to both the combustion air blower and the oil pump by means of a connector. It also has a thermal overload that is part of the assembly. The motor should require only minor maintenance, which includes a few drops of oil on an annual basis.

If the burner motor does not operate, follow these troubleshooting points:

1. Depress the reset button. If this action starts the motor, then continue with the rest of the steps to find out why it shut off on thermal overload.
2. Check that the supply voltage is between 110 V and 120 V.
3. Lubricate the pump with a few drops of SAE 20 nondetergent oil if it has not been lubricated recently.
4. Check the pump to make sure it is not creating a drag on the motor. If it is, disengage the pump and rotate the shaft of the pump. Replace the pump if it is hard to turn.
5. Check that the combustion blower rotates easily and is not bound.
6. Rotate the motor shaft by hand. If it turns easily but still will not operate electrically, replace the motor.

COMBUSTION AIR BLOWER

All oil burners have a combustion air blower. For complete combustion to take place, a blower is needed to deliver air to the flame. One gallon of oil requires 1,540 ft^3 of air in order to burn. The burner cannot mix all the air with the oil as approximately 20 to 30 percent excess air is required. Thus, most burners furnish between 2,000-2,200 ft^3 of air/gallon of oil.

The amount of air delivered by the combustion air blower determines the capacity of the burner. And, the amount of air delivered to the nozzle tip and mixed with the oil determines how much oil is burned. The nozzle receives the air and mixes it with the oil, and it is desirable to have turbulence at the nozzle. This reduces the excess air needed for the mixing process.

To provide this turbulence, burners have a bladed diffuser or stabilizer at the end of the blast tube. The diffuser increases the velocity of the air by restricting the exit point, which results in a better mixing capability. The air control shutter determines the amount of air delivered by the combustion air blower.

One common type of air shutter consists of slotted rings, one of which fits over the other. The outer ring is held in place by a locking screw. Loosening this screw and rotating the outer ring allows the service technician to expose more of the open port area and increase the amount of combustion air to the burner. Rotating in the opposite direction closes the ports, reducing the amount of air.

IGNITION

Most high-pressure, gun-type burners use electric ignition, consisting of a step-up transformer, high tension leads, electrodes and ceramic insulators.

Electrodes. Ignition is accomplished by jumping a spark (created by the 10,000-V secondary of the transformer) across a gap between a pair of electrodes located over and slightly ahead of the end of the nozzle, Figure 6-9. The gap between the electrodes and their location, both in terms of height above the centerline of the nozzle and the distance in front of it, is quite important. As a general guide for residential burners, the spark gap is fairly standard and should be 3/16 inch, unless the manufacturer's directions specify otherwise. Distance from the electrode tips to the centerline of the nozzle is 1/2 inch for nozzles with spray angles over 45 degrees. This is reduced by 7/16 inch for a nozzle with spray angles at 45 degrees and 3/8 inch for a nozzle with spray angles at 30 degrees. The distance from the electrode tip to the nozzle center varies according to the spray angle size, but the variation is not great (usually 5/16 inch). To prevent the spark from arcing and shorting out, electrodes should be at least 1/4 inch from any metal part in the burner.

Special porcelain insulators, which are capable of withstanding the voltages produced by the secondary of the transformer, hold the electrodes. Clips or other devices hold the insulators, which usually connect to the oil line. Special mountings secure the oil line in the blast tube and keep it in place. The back end of the electrode has buss bar contacts or insulated leads which complete the circuit to the secondary terminals of the transformer.

Checking the Electrodes. An ignition that is defective or delayed can create major and dangerous problems resulting in blow-back of the flame, therefore the electrodes need to be checked.

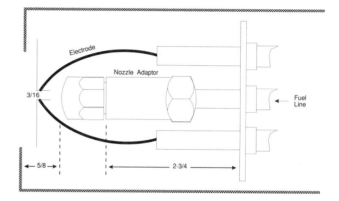

Figure 6-9. *Electrode Spacing Diagram*

The high voltage produced by the transformer takes the path of least resistance. Cracked or damaged electrode insulators cause the spark to jump to other metal parts in the gun-type burner. For this reason, the insulators on the electrodes should be removed and carefully examined for any cracks, breaks or darkened surfaces which would indicate leakage of electricity.

To prevent the spark from jumping to adjacent metal parts, no part of the electrodes should be closer than 1/4 inch to other metal parts in the burner. The electrodes should be out of the direct path of the oil spray, as oil can cause carbon build up across the gap, shorting out the electrodes and killing the spark. The electrodes should be visually checked to make sure there is no carbon build up. If carbon build up exists, a file may be used to clean the electrode tips so they do not resist the spark.

A poor, smoky fire caused by lack of draft or combustion air can lead to carbon or darkened spots on the insulators. This should be corrected. Carbon build up causes insulators to act as conductors, and the possibility of a short to some other metal part then exists. Also, carbon and soot build up rapidly if there is not enough spark for good ignition and combustion.

Electrodes should be checked to make sure that the ends connecting with the ignition transformer make good contact. If good contact does not exist, the following steps should be performed:

1. Bend or form the buss bar or plate to make good contact.
2. Determine the contact by feeling the resistance while the transformer is being secured in place.
3. Coat the buss bars with a tracing material and close the door if doubt exists as to whether or not the electrodes are making good contact with the transformer.
4. Observe contact marks, as these show whether or not firm contact is made.

When replacing the burner tube and ignition assembly, the burner tube, when reassembled, should not extend into the combustion chamber. The outer edge of the burner tube should be exactly flush with the inside of the combustion chamber. On a circular combustion chamber, this means that the center of the blast tube will be exactly on the inside diameter of the combustion chamber and that the edges will be slightly back from it. On a square chamber, the entire face of the burner tube will be exactly square with the combustion chamber, Figure 6-10.

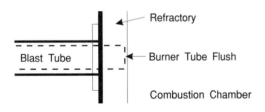

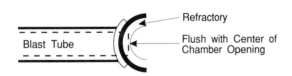

Figure 6-10. *Blast Tube Position in Combustion Chamber*

Transformer. The primary of the transformer is 120 V and the secondary is 10,000 V. The voltage produced is about 23 milliamps (0.023 A), so due to the low amperage, it is not particularly dangerous. A tar-like compound covers the coils of the transformer to protect them against corrosion and other external elements. A defective or weak transformer should be replaced, rather than repaired, at the job site.

Checking the Transformer. For good ignition to occur, the ignition transformer must be capable of producing

maximum spark. As there is a 10,000 V potential across the contacts of the transformer, caution should be exercised. The following steps should be performed when checking the transformer:

1. Tilt the transformer open to expose the terminals and apply power to the burner.
2. Take a well-insulated screwdriver and lay it across the transformer terminals to create an arc.
3. Draw the arc as far as possible. It should be between 1/2 and 3/4 inch in length.
4. Replace the transformer if the spark is weak or does not draw to this distance, there is no spark, or if it is weak in one terminal and not the other.

OIL BURNING NOZZLES

Atomizing, fuel delivery, and patterning are the major functions of the oil burning nozzle, Figure 6-11. These are defined as follows:

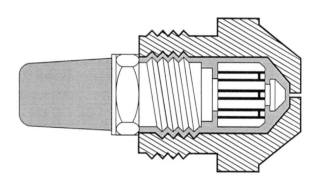

Figure 6-11. *Typical Oil Burner Nozzle*

- Atomizing, as stated previously, concerns breaking the oil into tiny droplets. This is necessary, because oil must be atomized before it can burn. It is then important to establish the flame front very close to the burner head.
- Fuel delivery must be consistent, as the nozzle is designed to deliver a fixed amount of fuel to the combustion chamber. This requires nozzles in a wide variety of flow rates to match the various manufacturer's furnace designs.
- Patterning is a uniform spray pattern the nozzle uses to deliver the atomized fuel to the combustion chamber. The nozzle must also employ the spray angle best suited to the requirements of that particular burner. Spray patterns range from a hollow cone, to a solid cone, to what is also called an all purpose spray pattern, Figure 6-12. Common spray angles are: 30, 45, 60, 70, 80 and 90 degrees. Normally, spray angles from 70 to 90 degrees are used in a square or round combustion chamber, Figure 6-13. Spray angles of 30 to 60 degrees are used in rectangular or long, round chambers, Figure 6-14.

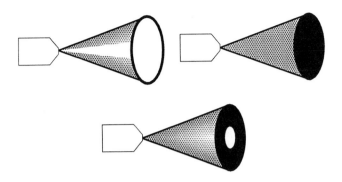

Figure 6-12. *Typical Nozzle Spray Patterns*

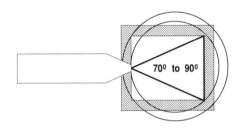

Figure 6-13. *Square or Round Combustion Chamber Spray Angles*

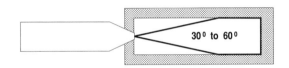

Figure 6-14. *Rectangular or Cylindrical Combustion Chamber Spray Angles*

The nozzle includes a strainer, as dirt can clog the nozzle and cause improper combustion. In spite of the strainer, some manufacturers recommend changing the nozzle once a year, anyway. This is due to oil erosion, which can sometimes change the orifice size and the spray angle into the combustion chamber. Consequently, the nozzle should be checked periodically for wear and changed when wear seems excessive. Special nozzle wrenches are available for this purpose. As a new nozzle has a body made of brass, it must be handled very carefully.

Nozzle manufacturers use different methods to measure spray angles and patterns. However, the manufacturers do supply a chart showing the proper settings. This chart should be considered a guide only, as the information is subject to change. For the most part, though, the nozzles give equal performance.

FUEL UNITS

The nozzles receive oil from the oil tank at a constant operating pressure (100 psi) due to the fuel unit. The fuel unit consists of the oil pump, a pressure-regulating valve, and a strainer. Pumps are made up of gears and can be

single stage or two stage. In both types of pumps, a coupling connects the pump shaft to the motor shaft, so the pump operates at a constant speed; its speed matches the motor.

Single-Stage Pumps. Single-stage pumps have a single set of gears and are used when the fuel oil tank is stored above the burner. This way, the fuel oil can flow to the burner by way of gravity. Single-stage pumps can be used in conjunction with a single-pipe supply system, which means there is only one pipe connecting the tank to the burner. However, single-stage pumps can also be used with a two-pipe supply system. Figure 6-15 shows a single-stage pump circuit diagram.

Two-Stage Pumps. When the fuel oil tank is stored below the burner, a two-stage pump is used, Figure 6-16. As the name implies, these pumps have two sets of gears: one set to lift the oil to the low-pressure, or inlet, side of the pump, and the other set to supply the oil, under pressure, to the nozzle. Two-stage pumps work in the same manner as the one-stage pumps, however, the extra gear, called a suction pump gear, produces a greater vacuum for purposes of lifting the oil. A two-stage pump requires a two-pipe supply system.

Both types of oil pumps are designed to pump too much oil to the nozzle. The nozzle cannot handle the extra oil, so the oil is sent to the low-pressure, or inlet, side of the pump in the single-pipe system and to the oil storage tank in the two-pipe system.

Both single- and two-stage pumps have a number of outlets for the required piping connections. The bypass plug is removed when the pump is shipped, so the pump is ready for single-pipe installation. The 1/16-inch bypass plug must be inserted (with a 3/16-inch Allen wrench) into the 1/4-inch return line port for two-pipe systems.

Pumps are normally not repaired by the service technician. Instead, the pump should be replaced if it is found to be defective.

Pressure-Regulating Valve. The pressure-regulating valve is designed to deliver the oil to the nozzle at a constant operating pressure of 100 psi. This pressure provides for a constant gallon/hour flow rate and spray pattern. The pressure-regulating valve also makes sure that the oil supply is cut off quickly when the burner goes off, so there is no after-drip of oil while the burner is off.

Strainer. The strainer is placed ahead of the pump and ensures that no dirt or foreign material enters the pump. The pump and the strainer should be checked and cleaned with clear fuel oil or gasoline on periodic maintenance inspections. A new cover gasket should be installed after each cleaning.

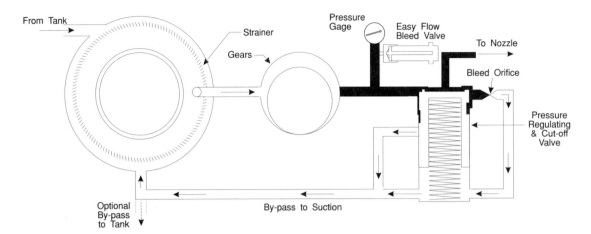

Figure 6-15. *Single-Stage Pump Circuit Diagram*

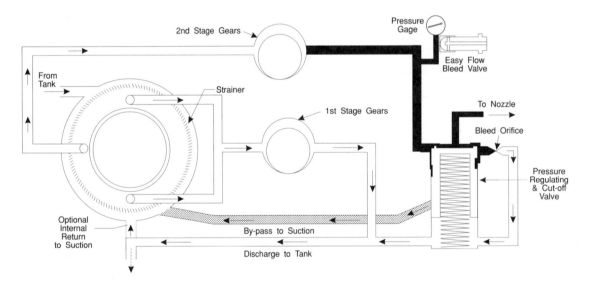

Figure 6-16. *Two-Stage Pump Circuit Diagram*

AIR LEAKS

All valves, filters, splices and other fittings should be checked for air tightness, as air leaks are often to blame for service problems. A small air leak can cause the oil to drip into the combustion chamber after the burner has shut down. This is called afterfire. Afterfire fouls the nozzle, causes odors and can also cause unusual pump noises.

PURGING THE FUEL LINE

After checking the fittings for tightness, the fuel line can be purged of any trapped air in the following manner:

1. Turn off the disconnect switch to the oil furnace and remove the wires from the flame detection (FF) and thermostat terminals (TT) .

2. Jumper the thermostat terminals.

3. Loosen the pump bleed valve by about one-half turn and slip a 1/4-inch inside diameter (id) hose over this valve. Place the other end in a large container.

4. Close the disconnect switch, which starts the burner, and jumper the FF terminals within 45 seconds so that the burner operates continuously. Be sure the burner safety is locked out and that fuel ignition occurs before continuing the procedure.

5. Collect at least a pint of oil, or continue until the oil is clear and free of bubbles before stopping the bleed procedure.

6. Two-stage pumps using two-pipe systems are self-venting, which means they do not have to be bled, but if it is desirable to speed up the process, the above procedures can be followed.

TESTING FOR LEAKS BY CHECKING PUMP PRESSURE

Pump pressure and vacuum should be checked for air leaks. To check this pressure, follow these four steps:

1. Remove the plug from the pump port marked Gage, and connect a pressure gage (0-300 psi) to the port.
2. Jumper the TT terminals of the primary control to turn the burner on. The pressure in the oil pump should rise to 100 psig, plus or minus 5 psig. If it is above or below this, adjust the pump pressure by turning the adjustment screw on the pressure regulator. Turning the regulator clockwise increases the pressure.
3. Cycle the burner two or three times with the TT jumper wire by removing one end, stopping the burner and then reinserting it after a wait of 2 or 3 minutes. Make sure the pump pressure remains at 100 psig, and the burner continues to operate correctly. If not, bleed and adjust the pump pressure again.
4. Remove the jumper wire and gage and reinstall the port plug. Remove the bleed hose.

TESTING FOR LEAKS BY CHECKING PUMP VACUUM

Leaks in the pump and oil supply line can be detected by checking to see if a pump draws the proper vacuum. The following steps should be performed when testing for leaks:

1. Remove the bypass plug and return line fitting, if necessary, in order to set up the pump for a single-line operation.
2. Install the vacuum gage in the unused intake port.
3. Start the burner and open the bleed valve to make sure there is oil in the pump. Stop the burner, remove the oil line from the intake elbow and cap the elbow.
4. Start the burner, closing the disconnect switch, and open the bleed valve until the vacuum gage registers approximately 10 inches of vacuum.
5. Close the bleed valve and stop the burner. The vacuum gage reading will drop slightly until the check valve closes. The vacuum gage reading should now hold constant. If the gage reading continues to drop after shutdown, check all plugs, nuts and caps covering the pressure adjustment screw for tightness.
6. Repeat steps 3, 4 and 5; if the pump still does not hold a vacuum, replace it.
7. Reinstall the oil line to the tank and run the burner to obtain a running vacuum reading. This reading should be approximately 1 inch of vacuum for each foot of lift. A reading substantially higher than this indicates a restriction in the line fittings and possibly the filter. This completes the series of tests.

COMBUSTION EFFICIENCY

To ensure maximum combustion efficiency in a high-pressure burner, it is necessary for the service technician to run several tests using combustion test equipment. These tests include draft, net stack temperature, carbon dioxide, and smoke tests. The service technician then makes adjustments, as necessary, so near-perfect combustion can be reached.

PREPARING FOR COMBUSTION TESTING

To be effective, combustion testing must follow an orderly and planned procedure so that the variables can be tested in the proper sequence. Most service technicians eventually develop their own procedures when preparing and testing for combustion efficiency, however, the following steps are usually employed.

1. Open the main burner switch. This switch will be used to start or stop the burner since it is located close to the burner and can be quickly switched off if there is a problem on initial start. Check the fuses in this switch, particularly if the customer complaint was concerning no heat.
2. Inspect the combustion chamber and clean out any accumulated oil. Leaving the observation port open usually burns out any oil absorbed in the combustion chamber.
3. Raise the thermostat temperature 5 to 10 °F to ensure continuous burner operation.
4. Close the remote control burner switch (if there is one).
5. Drill a 1/4-inch hole, if one does not already exist, in the flue between the flue collar of the furnace and the barometric regulator. This hole should be at least 6 inches from the regulator. A 1/4-inch hole in the combustion area is also necessary; this might be provided by removing a bolt in the inspection door. On other units, the air louver may be opened slightly to provide a place for the draft tube to be inserted. Seal off balance with paper or cardboard.
6. Insert a stack temperature thermometer (200 to 1000 °F range) through the 1/4-inch hole in the flue to make sure the end of the thermometer stem reaches into the center third of the flue pipe diameter.
7. Open the inspection port or door and adjust the flame mirror so that the flame can be easily observed. Start the burner by closing the main burner switch. If the unit fails to ignite, and if there is no oil mist in the chamber, suspect oil stoppage. If there is oil mist in the chamber, suspect However if there is oil mist in the chamber, suspect a failure in the ignition circuit.

INITIAL INSPECTION

An overall inspection of the combustion chamber should be made before starting any testing. The following steps include some of the elements a service technician should check:

1. Note whether burner ignition is instant or delayed.
2. Check the flame for one or more colors. The flame may be orange, yellow, white, or a combination of these colors and may also show sparks within it. The desirable color is orange with some yellow in it.
3. Inspect the flame shape. The flame shape should be uniform and cover a broad area of the combustion chamber. The ideal flame is a sunflower shape and evenly covers a broad area. A lopsided flame only covers areas in the chamber not being exposed to the flame. A poor or unbalanced flame usually indicates a bad nozzle which may have to be replaced.
4. Check for flame impingement and note where it occurs (i.e., bottom, sides or rear of the combustion chamber). This can be a sign of a bad nozzle or incorrect size nozzle.
5. Check for odor near the burner, at the observation door, and at the draft regulator. No odor should exist.
6. Note and record whether there are any rattles, hums or pulsations on ignition and if so, whether they occur during starting or running conditions.
7. Look for soot deposits at the flue combustion chamber within the furnace.

TESTING FOR DRAFT

In order for a burner to operate efficiently, a proper draft is necessary. When testing for draft, the actual values measured must be recorded in order to use them for future calculations and references.

When testing for draft, perform the following steps:

1. Zero-out the draft gage. Make sure this is done on a level surface.
2. Turn on the burner and let it run for at least 5 minutes before testing.
3. Insert the draft tube into the hole made previously in the combustion area. This provides a means to check the overfire draft. The draft gage measures in inches of water (W.C.). The overfire draft should be 2 inches W.C. in order to keep rumblings, smoke, and odor from the basement.
4. Seal the tube with a piece of sheet metal or tape so no air leakage exists.
5. Insert the draft tube into the flue pipe to check the flue draft. The flue draft should be between 4 and 6 inches W.C.
6. Record both draft readings.

TESTING NET STACK TEMPERATURE

The net stack temperature is an important indicator of efficient furnace operation. Overfiring, lack of baffling, or a dirty heat exchanger are just a few of the problems a high net stack temperature can indicate.

Perform the following steps in order to test the net stack temperature:

1. Run the burner for approximately 5 minutes, or until a constant temperature is reached.
2. Insert the thermometer in the hole drilled in the flue. Record the flue temperature. It should be approximately 500 °F. A higher reading indicates that excess heat transfer in the heat exchanger is inefficient. As stack temperature comes down, the efficiency of the furnace increases, as there is loss of heat from combustion.
3. Record the basement air temperature in degrees Fahrenheit.
4. Calculate the net stack temperature in degrees Fahrenheit by subtracting the basement temperature from the flue temperature.
5. Remove the thermometer from the flue hole.
6. Compare reading with manufacturer's specifications.

TESTING FOR CARBON DIOXIDE (CO_2)

Testing for CO_2 is important when checking combustion efficiency. A low CO_2 reading can signal a clogged nozzle, excess combustion air, or an incorrectly-set oil-pressure regulator. In this test, the same hole used for stack temperature an flue draft is used. In order to test for CO_2, perform the following steps:

1. Run the burner for at least 5 minutes.
2. Wait for the temperature to stop rising (this can be done by inserting a thermometer).
3. Check the CO_2 equipment. There are two methods used to check the equipment.
   a. Squeeze the aspirator bulb, place a finger over the metal end of the sampling tube, then release the aspirator bulb. The aspirator bulb should remain deflated. If it does not, this indicates the check valve on the rubber cap and the aspirator bulb are bad. These must be replaced before the CO_2 analyzer can be used.
   b. Place a finger tightly over the rubber cap end of the tube and squeeze the aspirator bulb. You should not be able to depress the bulb. If it can be depressed, the check valve on the metal tube end of the aspirator is bad and must be replaced.
4. Insert the CO_2 sampling tube into the hole in the flue pipe. Note that the metal tube has a small U-shaped end to help hold it in place in the flue.

5. Depress the plunger on top of the CO_2 analyzer to release any flue samples that might remain from previous checks.
6. Following the procedure outlined by the test instrument manufacturer, remove a test sampling.
7. Mix the test fluid in the instrument with the sample gases in the flue.
8. Read and record the percent CO_2 as indicated on the instrument scale. The CO_2 reading should be between 8 and 10 percent.

TESTING FOR SMOKE

This test determines whether the flame is smoking excessively, causing incomplete combustion and soot build-up in the furnace. Soot build-up can result in wasted fuel and reduced heat transfer. This excessive smoke can also indicate an improper fuel-air ratio, a poor fuel supply, or an improper fan collar setting. Measurements are taken by infusing filter paper with a certain number of smoke-laden flue products.

To perform a smoke test, follow these steps:

1. Run the burner for at least 5 minutes before testing.
2. Make sure there is filter paper in the tester. If there is no filter paper, loosen the front end of the tester slightly and insert filter paper into the holding slot. Retighten the tester.
3. Insert the free end of the sampling tube into the hole in the flue pipe used in the other tests.
4. Pull the smoke tester handle, which is similar to a small bicycle pump, about 10 times. Use slow, even strokes and rest several seconds between each complete stroke.
4. Remove the filter paper.
5. Compare the filter paper with the scale provided with the instrument.
6. Record the reading.

Perform the following steps if the reading shows excessive smoke, according to manufacturer literature:

1. Increase the amount of combustion air by loosening the setscrew and rotating the outer ring to expose more port area.
2. Operate the burner for 4 or 5 minutes and recheck the CO_2. If the CO_2 is less than 8 percent and the smoke test is normal, reduce the combustion air and recheck both. If the CO_2 is less than 8 percent and the smoke is above normal, suspect leakage of air at the inspection door, flue pipe or burner receiving tube. If there is no leakage at these points, the nozzle may need replacement.

CALCULATING COMBUSTION EFFICIENCY

There are simple combustion efficiency slide rules which directly determine combustion efficiency from the previous CO_2 and net stack temperature tests. There are also electronic combustion analyzers which can save time by taking the same tests as previously outlined, but these devices save time and are easier to use, due to a convenient digital readout.

FIELD WIRING

The first step in field wiring is to run the 120-V, two-wire service from the main breaker panel in the house. The size of the main breaker or fuse and the size of the wire are specified in the instructions that come with the unit, and are in accordance with the National Electric Code. Most areas require conduit for all parts of this service.

A disconnect switch must be provided on or near the furnace. This is fused according to the manufacturer's instructions. The two wires are connected to the line side of the box with the fuse in the hot leg (black wire). Two wires are run from the disconnect to the line voltage make-up box in the furnace. The black wire is connected to the wire going to the fan-limit control. The white wire is connected to the white lead from the primary and the white lead to the blower.

If the oil burner assembly is shipped separately from the furnace, or if its leads are not connected, then the black wire from the burner attaches to the orange wire at the primary control. The white wire from the burner connects to the white wire at the primary.

Two low voltage wires (R and W) run from the thermostat and connect to T and T at the primary. Cad cell leads from the burner assembly are connected to terminals S and S (or FD and FD) at the primary. If fan control at the thermostat is desired, then G from the thermostat is connected to G at the terminal block or fan control in the furnace. This completes the field wiring for a heating-only oil furnace.

REVIEW QUESTIONS

1. What is the fuel most used in oil burners, and what is its Btu output?
2. What is a primary control?
3. When the burner locks out on safety, what should be checked?
4. What is the purpose of the flame detector?
5. What should be the normal cad cell resistance with the burner running?
6. Name the components of the high-pressure burner.
7. When a burner motor goes off on overload, what would be some of the causes?
8. What purpose does the combustion air blower serve?
9. What is the voltage of an ignition transformer?
10. Where are electrodes positioned?
11. Explain how to check an ignition transformer.
12. In the combustion chamber how is the burner tube positioned?
13. Explain the functions of the burner nozzle.
14. What are common spray angles?
15. Describe the oil pump's function.
16. Where are single-stage pumps used?
17. Where are two-stage pumps used?
18. What is the pressure-regulating valve used for?
19. What is a proper operating pump pressure?
20. What is a normal stack or flue gas temperature, and what CO_2 reading is desirable?

Chapter 7
Electric Heating

Electric heat has been around for some time, in the form of heat elements embedded in the ceiling or floor, or as electric convection units and cable heat. Consumers like electric heating systems, because no vent pipes are needed (as with gas and oil) to remove waste; the cost of the actual electrical heating system is usually lower than other systems; and electric heat is efficient, as there is very little loss of electrical energy. The one drawback, however, is that electric heat can be very expensive, because it takes a large amount of electrical energy to produce heat.

Applications of electric heat may vary. For example, an electric furnace that includes an air conditioning coil is known as a fan coil unit. In this unit, the cabinet contains a blower, electric heater and cooling coil, and the electric heat is placed on the downstream side of the blower, Figure 7-1. As another variation, when a unit is used as the indoor section of a split heat pump, it is called a heat pump fan coil. Electric heat elements can also be inserted into the duct system to provide additive heat for add-on rooms or hard-to-heat rooms. Or, an entire house can be heated by elements remote from the basic air mover. Heating elements used in this manner are called duct heaters, Figure 7-2.

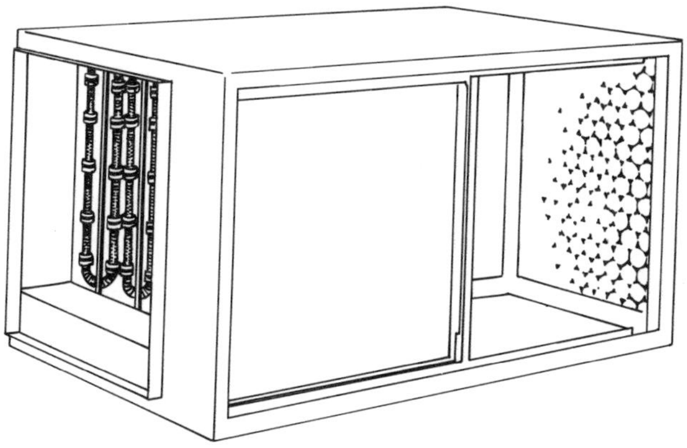

Figure 7-1. *Electric Heater on Downstream Side of Blower*

ELECTRIC FURNACES

Electric furnaces are similar to gas or oil furnaces in that they all use the same type of duct system, and the cabinets, filters, blowers and motor options are the same. One extra requirement is that better insulation is needed in houses and businesses equipped with electric furnaces, in order to cut electric consumption and costs to a minimum.

HEATING ELEMENTS

An electric furnace differs from oil or gas furnaces in many ways. The major difference is that there is no combustion in an electric furnace, so there is no need for a heat exchanger. Instead, a resistance heating element produces the heat. This heating element consists of wound, 20-gage, nickel-chromium (nichrome) wire mounted on ceramic or mica insulation. Nichrome is used, as the high nickel content slows oxidation and separation of the wire. The heating element is placed in the air stream, so the heat it produces can be transferred directly to the air, without an intermediate heat exchanger.

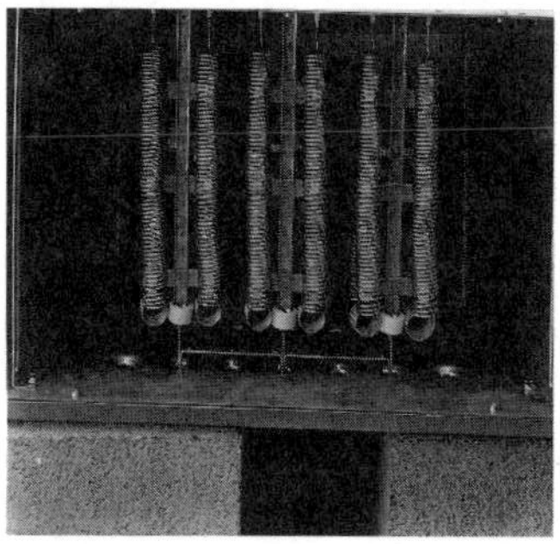

Figure 7-2. *Electric Heating Element. Courtesy, Carrier Corporation, a Subsidiary of United Technologies Corporation.*

Electric furnaces also do not require access spaces in the cabinet for combustion air, and flues are not required, because there are no products of combustion to remove. As there are no flue or combustion air losses, electric furnaces are considered to be 100 percent efficient. Consequently, the heat input and heat output is the same; for each kilowatt (kW) of electricity the furnace receives, 3,415 Btu of heat energy is produced.

Electric heat elements may be used individually, or in multiples in order to produce the required heat output. A single element has an input capacity between 4.5 and 6.0 kW input, or between 15,000 and 20,000 Btu of heat output. When two or more heating elements are used, they are usually step-started. This means that the second element does not turn on until the first element produces heat, the third element does not turn on until the second element produces heat, and so on. There are as many steps as there are elements, and stepping continues until all elements are producing heat. Stepping avoids putting a high-power load on the line at one time and overloading the circuits when starting.

The elements are not position sensitive, so the same ones can be used in an upflow, downflow or horizontal furnace. Each element is on a separate branch circuit with fuses in both legs of the line. Elements can be checked for proper operation by pulling out the bracket and checking for any broken or loose wires. If there are no visible defects, a continuity check can be made with an ohmmeter or test light.

Multiple elements start in sequence, with a slight time delay between each element coming on the line. This can be easily checked with an ammeter, set on the highest current range, attached to the main load wires of the circuit. As each element comes on, there is a very definite jump in current indicated on the ammeter. The service technician can determine whether or not all elements have sequenced on in proper order by counting the surges as the elements come on the line. If all of the elements do not come on in sequence, it is possible that one stage is bad, or the outdoor thermostat (if the unit has one) may be holding off the last one or two elements. If this happens, the outdoor thermostat can be reset above the outside temperature. This will allow the final elements to come on the line in sequence.

LIMIT CONTROLS

An electric furnace has a limit control for each element. A limit control is a temperature-activated switch that cuts off the element if the surrounding air temperature becomes too high, Figure 7-3. This cut-off point varies, but most range between 160 and 170 °F. The manufacturer determines the cut-off point, and the setting is nonadjustable. The limit automatically resets when the ambient temperature decreases.

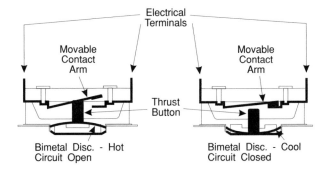

Figure 7-3. Limit Control

The limit control mounts in the front bracket plate and senses both the temperature in the compartment and the radiant effect of the element. As each element has its own temperature-sensing limit control, the limit interrupts power to its element only, not disturbing the other elements. Electric furnaces, therefore, do not require secondary limits for horizontal or downflow units, as are found in oil and gas furnaces.

To check the limit controls, follow these steps:

1. Either disconnect the blower motor or block the return air.
2. Place a thermometer in the cabinet. When the temperature reaches 170 °F, or the designated cut-off point, the limit should open and cut off power. There should not be much variance in this cut-off point.
3. Shut off the furnace and replace the limit if the temperature rises 10 to 15 °F over the cut-off point and the limit does not open.

THERMAL FUSES

The thermal fuse, or fusible link, is another safety device found on an electric furnace. The thermal fuse acts as a backup limit control. It is set to open at a temperature considerably higher than the limit, typically 300 to 310 °F. In the event that the limit does not open on an excessive temperature rise, the thermal fuse melts, opening the circuit.

The thermal fuse is in series with the limit and the heating element. Unlike the limit, if the thermal fuse opens or melts, it does not automatically reset. Some must be manually reset, and others replaced. In older furnaces, it is often necessary to replace the entire element if the thermal fuse opens.

CAUTION: There is no valid service check for proper operation of the limit control on older units with nonreplaceable thermal fuses. If the blower belts are removed or direct drive leads disconnected (as in a gas furnace limit check), the temperature exceeds the limit control point and burns out the fusible plug, requiring changing of the entire element. Consequently, a thermometer should be used to check the limit. If the limit is faulty, the furnace can be turned off before the thermal fuse melts, as there usually is a wide differential (130 to 140 °F) between these points.

BLOWER AND ELEMENT SEQUENCING

Electric furnaces heat up rather rapidly, so it is desirable to have the blower turn on at the same time as, or slightly ahead of, the first heater element.

A sequencer can be used to energize the blower and the first element simultaneously, or with a time delay of about 30 seconds (this delay period can vary according to the control used). The sequencer then energizes the next element 20 seconds later, and so forth. Shutoff occurs (after a 3-minute delay) in the reverse order.

Two time-delay relays are used to bring the blower on slightly before the heater element. The first relay, once energized, turns on the blower after a time delay of approximately 15 seconds. The second relay, energized at the same time, waits 20 to 25 seconds before energizing the first heating element. Additional relays, with longer time delays, can be used to turn on additional elements. This system then shuts down in reverse order, keeping the blower operating until all elements are off.

Another sequencing system includes a set of auxiliary, low-voltage contacts with each time-delay relay. When the first relay is energized, the auxiliary contacts close, also energizing the heater in the second relay. When this relay is energized, it brings on the second element and energizes the heater in the third relay, and so forth. This continues for as many elements as are in the system.

To check sequencer operation, perform the following steps:

1. Set the thermostat to call for heat.
2. Place the fan selector switch on automatic.
3. Place the voltmeter leads on terminals C and W1 on the low-voltage terminal block. The meter should show 24 V.
4. Place the leads on each of the high-voltage terminals. A voltage reading here indicates that the contacts are open. They should make within 30 seconds (or whatever the time delay period is) and show no voltage.
5. Turn off the unit and, when the contacts open, look for the voltage reading on the meter.

HEAT ANTICIPATION

There is only one major difference in the way heat anticipation is checked in an electric furnace as opposed to a gas or oil furnace; otherwise, an electric furnace is checked in exactly the same manner. The difference is that the heat anticipation setting is usually at a considerably lower point (0.25 to 0.30 A) in an electric furnace than it is in a gas or oil furnace. However, the heat anticipator can exceed 1 A when several elements are staged with the heaters in series.

To set the heat anticipator in a single-stage thermostat, perform the following steps:

1. Move the thermostat to its lowest setting.
2. Attach one clip of the amperage multiplier to the heat anticipator lever, and attach the other clip to the common terminal of the wires to the mercury bulbs.
3. Take an amperage reading after a minute or two of operation. Divide this reading by 10 to convert it into actual amperage. Set the heat anticipator lever at this point, Figure 7-4.

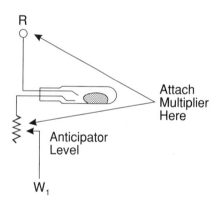

Figure 7-4. Single-Stage Heat Anticipator

To set the heat anticipator in a two-stage thermostat, perform the following steps:

1. Check the first stage as if it were a single-stage thermostat, and set the heat anticipator lever to the reading obtained.
2. To check the second stage, place a jumper wire across the previously used common terminal and the first-stage anticipator lever. Then, attach the amperage multiplier to the second-stage heat anticipator lever and the other side to the common terminal, with the wires running to the mercury bulbs, Figure 7-5.
3. Take the amperage reading after a minute or two of operation. Divide this reading by 10 to calculate the actual amperage. Set the amperage on the second-stage heat anticipator.

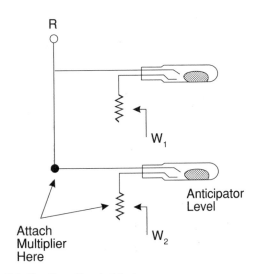

Figure 7-5. Two-Stage Heat Anticipator

TEMPERATURE RISE

The temperature rise across the electric heat elements is less than that considered normal in a gas or oil furnace, because an electric furnace converts 100 percent of its energy input into heat energy. In a gas or oil furnace, the temperature rise across the heat exchanger is between 80 and 100 °F, and is normally held at about 90 °F. In an electric furnace, the temperature rise is usually between 50 and 70 °F.

To check the temperature rise across an electric furnace, perform the following steps:

1. Place one thermometer in the return air plenum, close to the furnace, and a second thermometer in the supply plenum, just out of sight of the electric element, Figure 7-6.
2. Read each thermometer after the furnace has operated for several minutes, noting the difference between them.
3. Check manufacturer's literature for the acceptable temperature rise, as some furnaces can run as high as 85 °F
4. Slow down the blower, as described in Chapter 5 (Gas Heating), if the temperature difference is less than 50 °F, as this indicates the blower is running too fast.
5. Speed up the blower if the temperature difference is more than 70 °F, as this indicates the blower is running too slow.

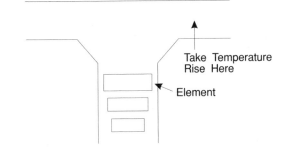

Figure 7-6. Temperature Rise Check

ELECTRIC FURNACE MAINTENANCE

Maintenance of an electric furnace is relatively simple mainly because it contains no major moving parts other than the blower. Also, an electric furnace has no flue or vent connections. The only major maintenance required is to check all fuses and electrical circuits for tightness. It is also necessary to check the blower and filter periodically (as described in Chapter 5, Gas Heating).

FIELD WIRING

An electric furnace requires either 240-V or 208-V, single-phase service. Therefore, a three-wire line voltage supply run from the main panel in the house to the furnace is needed. The disconnect and the fuse in the main supply are sized in accordance with the manufacturer's instructions which come with the unit. The wire size is also specified in the instructions, in accordance with the National Electric Code.

One circuit is permissible to supply power from the main supply to the disconnect. The disconnect switch, Figure 7-7 must be provided at the furnace. In addition, it must be located within sight of the furnace or be capable of being locked in the open position.

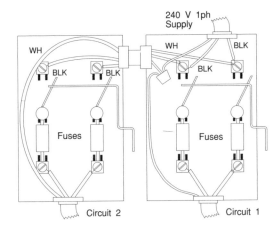

Figure 7-7. Field Wiring for a Pair of Disconnects

If the furnace does not draw over 48 A, then only one disconnect is required. This disconnect cannot be fused for more than 60 A. If the furnace requires greater amperage, additional disconnects must be supplied—one for each circuit, not exceeding 48 A. These circuits must be enclosed in conduit, and watertight strain reliefs or connectors are required at both ends. Each hot (black) wire in the disconnect must have a fuse. These fuses can be either the standard cartridge-type or circuit breakers.

The National Electric Code also requires that these circuits be grounded at the furnace. Most furnaces provide a ground connection in the control box adjacent to the high-voltage terminals, Figure 7-8. The white wire is grounded at this point and the two black wires are connected to the line voltage terminals in the furnace.

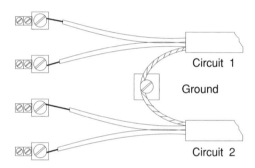

Figure 7-8. *Control Box Grounding Circuits*

Two low-voltage wires (three with fan control and four with two-stage heat plus fan control) run from the thermostat to the furnace. Most modern furnaces have a terminal block or fan control center marked in the same manner as the thermostat for the low voltage leads. Thus R, Wl, W2 and G from the thermostat connect to the R, Wl, W2 and G terminals in the furnace. This completes the field wiring.

GENERAL FURNACE MAINTENANCE

Service technicians should perform maintenance checks in order to prevent major service problems. Making these checks can also correct or eliminate the majority of minor service problems before they result in furnace down time. The following list includes basic maintenance procedures that can be performed on all types of furnaces:

1. Check all line voltage wiring and connections for tightness, good insulation and proper contact.
2. Check all low voltage wiring and connections for tightness, good insulation and good contact.
3. Change filters every 6 months, or more often if they are excessively dirty.
4. Check blower belts for cracks, wear or excessive slippage.
5. Check pulleys and drive setscrews for tightness.
6. Check pulleys and drives for proper alignment, and adjust belt tension if necessary.
7. Lubricate all motors and bearings per the manufacturer's instructions and make sure they are free and completely unobstructed.
8. Check the limit control for proper operation.
9. Check temperature rise through the furnace to see that the blowers are running at the proper speeds and the furnace is functioning as efficiently as possible.
10. Check thermostats for level, and check heat anticipation for proper setting.
11. Check burners, heat exchangers and cabinet compartments for cleanliness and proper operation.
12. Check safety devices such as thermocouples, gas valves, safety dropouts and cad cells for proper operation.

REVIEW QUESTIONS

1. What are the basic components of an electric furnace?
2. How does an electric furnace differ from a gas or oil furnace?
3. How many Btu are produced with an input of 1 kW?
4. Are multiple elements started in sequence rather than coming on all at once?
5. Why is the limit control on an electric furnace different than other furnaces?
6. Is there a secondary limit for horizontal units?
7. What is the thermal fuse?
8. When does the blower start in an electric furnace?
9. What is the temperature rise for an electric furnace?
10. Name the basic maintenance items for all furnaces.

Chapter 8

Hydronic Heating

A hydronic heating system uses water or steam to carry heat through pipes to the areas requiring heat. No forced air is required in hydronic heating. When the boiler is not on in a hot water system, hot water remains throughout the system. Therefore, the living space remains consistently heated, with no hot or cold areas. In a steam system, steam heats quickly, so the living space heats quickly; however, when the heat circulation stops, the living space also cools down quickly.

HOT WATER SYSTEMS

In a hot water system, water is heated in a boiler and sent through pipes to a terminal unit (i.e., radiator or baseboard unit). The terminal unit transmits the heat from the water to the surrounding air in order to warm up the living space. The water then returns to the boiler.

When discussing hot water systems, it is necessary to distinguish between open and closed systems, as well as whether gravity or forced circulation is used. These are discussed later in this chapter.

The main components in the hot water system include a boiler, circulating pump, piping system, and terminal unit.

BOILERS

The primary objective of a boiler is to heat the water that is to be sent to warm a living space. Boilers can use an oil, gas, or electric heat source in order to heat the water, and some larger boilers even use two different heat sources. A hot water boiler maintains water temperatures between 120 and 210 °F.

A boiler comes equipped with certain features which not only prolong its lifetime but also protect the owner from hazardous conditions, Figure 8-1. These features include an expansion tank, air vents, pressure relief valves, limit controls, and similar components.

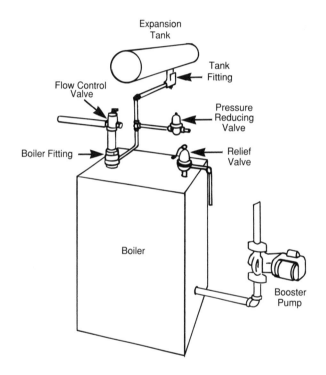

Figure 8-1. *Hot Water Boiler*

Expansion Tank. The expansion tank determines whether a system is open or closed. As water heats up, it expands. When the water expands, it compresses the air that is present in the system. The expansion tank collects or contains this compressed air. When the system is open, the expansion tank vents the compressed air into the atmosphere through air vents. In a closed system, the expansion tank performs the same action, however, it does not vent into the atmosphere. Once the system cools down, the water returns to its normal state, causing the compressed air in the expansion tank to also return to return to its normal pressure.

Air Vents. As just stated, air vents are required in an open system in order to vent the compressed air collected by the expansion tank into the atmosphere. Air vents also help rid the system of all other air. Air in the boiler at atmospheric temperature can cause corrosion, and as the temperature of the water increases in the system, this corrosion factor climbs even higher. Air pockets can also block the water circulation and cause undesirable noise. In an open system, this air is routed through the expansion tank; however, in a closed system, the expansion tank contains this air, rather than releasing it through air vents.

Pressure Relief Valve. The purpose of the pressure relief valve is to protect the boiler against excessive pressures. If boiler pressure exceeds the maximum allowable working pressure, the pressure relief valve opens, discharging the water until the boiler pressure returns to a safe operating level.

Limit Control or Aquastat. As in other heating systems, boilers contain a limit control, also called an aquastat. This aquastat is basically a sensory device that maintains a constant supply of hot water in the boiler while the system is not turned on. Normally, the aquastat consists of a thermostat bulb immersed in the boiler water. When the boiler water falls below the desired temperature setting, certain contact points in the aquastat close the electrical circuit. This starts the burner, which maintains the water temperature as set on the control. When the thermostat calls for heat, the thermostat contacts close, which completes the circuit to the 24-V coil in the aquastat. This causes the contacts to snap closed and complete the 120-V circuit to the water pump. The pump then continues to circulate water until the room thermostat temperature is satisfied. When the temperature of the boiler water exceeds the preset temperature range (about 190 °F), the aquastat shuts down the heating source.

In order to keep the system completely filled with water, the boiler must maintain a pressure higher than that set on the pressure reducing valve. This valve is located in the water feed line and is usually supplied with a city water pressure of about 65 psi.

CIRCULATING PUMP

The circulating pump, also called a centrifugal pump, is the heart of the closed hydronic system, Figure 8-2. These pumps move the hot water from the boiler through the piping, to the terminal units, and back to the boiler. Centrifugal force moves this water through the system, and this force depends on the speed of rotation of the impeller in the pump.

Figure 8-2. Circulating Pump. Courtesy, ITT Bell & Gossett.

The impeller is the part of the pump that rotates, thus forcing water through the system, Figure 8-3. It is very important that the impeller is forcing water in the proper direction. The impeller should rotate the water away from the center of rotation, thus forcing the water out into the system.

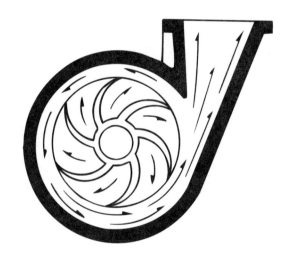

Figure 8-3. Impeller

PIPING SYSTEM

The piping system carries the hot water from the boiler to the terminal unit. There are several different pipe configurations that can be used, including one-pipe and two-pipe systems.

These one-pipe and two-pipe systems can operate either by gravity or force. Gravity circulation is caused by the difference in weight between the hot water in the supply branches and the cool water in the return branches. The weight of the hot water in the supply line is lighter than the return cool water. Therefore, when a supply branch is connected to the main line at a certain point, and the return is reconnected at the proper distance along the line, there is enough pressure difference between supply and return to cause water circulation.

Forced circulation is caused by a circulating pump (described previously). These pumps do not maintain pressure within the system; their only purpose is to keep water circulating throughout the system. This circulating pump is normally located near the boiler inlet on the return line.

One-Pipe System. As the name implies, this system is made up of one pipe which runs through the entire building, Figures 8-4 and 8-5. This one-pipe is also known as the main line, and supply branches come off this line, supplying hot water to each terminal unit. The piping in the supply branches is smaller in diameter than the main line, but the main line size remains constant from boiler outlet to boiler inlet. When the water is returned from the various terminal units, it flows through the main line back to the boiler.

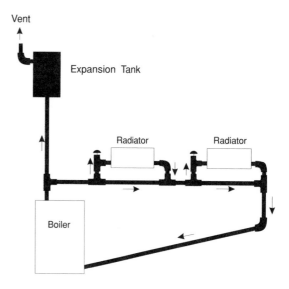

Figure 8-5. One-Pipe, Gravity-Circulation Hot Water System

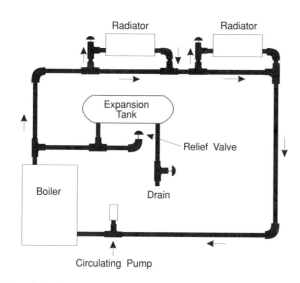

Figure 8-4. One-Pipe, Forced-Circulation Hot Water System

Two-Pipe System. This particular system consists of two main lines. One line is the supply line, supplying hot water to the terminal units, and the other line is the return line, which collects the cool water from the terminal units and returns it to the boiler. These two lines run parallel to each other with the supply main decreasing in size at each branch connection and the return main increasing in size at each branch connection, Figure 8-6. There are two types of two-pipe systems: reverse-return and direct-return.

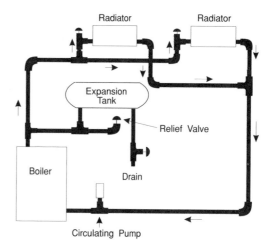

Figure 8-6. Two-Pipe, Forced-Circulation Hot Water System

Reverse-Return System. In this two-pipe system, the water supplied to the first terminal unit is the last water returned to the return line. Conversely, the water supplied to the last terminal unit is the first water returned. This system requires more pipe than the direct-return system, however, it provides better flow distribution and balance. The two-pipe, reverse-return system is often used in light commercial applications, however, it is rarely used in residences.

Direct-Return System. In contrast to the reverse-return system, the water supplied to the first terminal unit is the first water returned, Figure 8-7. The second terminal unit then returns its water after the first terminal unit, and so on down the line. This consecutive return of water in this system causes progressively greater friction loss, and the flow circuits become unbalanced. Therefore, the pipe sizes used in this system must be carefully selected to compensate for the friction loss at each circuit. As it is very difficult to balance this type of system, it is rarely used in residential applications.

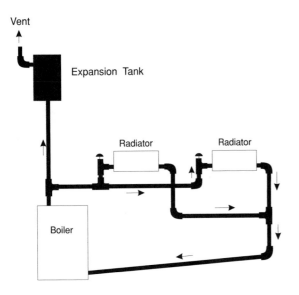

Figure 8-7. Two-Pipe, Direct-Return, Gravity-Circulation Hot Water System

Check Valves. The check valve is located before the boiler, and its job is to prevent the water from backing up in the system. When the system is on, the check valve is open, allowing water to pass through it. When the system is not running, the check valve closes, stopping water flow.

TERMINAL UNITS

There are several different types of terminal units, including radiators, baseboard units, air coils, submerged heaters, and heat exchangers. The application determines which type of terminal unit is used. The most widely-used terminal unit is the baseboard unit.

Baseboard Unit. Baseboard units contain finned tubes, Figure 8-8. Air enters the bottom of the unit and passes over these hot finned tubes. The warmed air then moves into the living space via a damper. These dampers control heat flow.

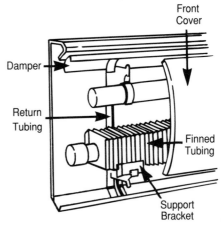

Figure 8-8. Two-Pipe Baseboard Unit

STEAM SYSTEMS

Steam systems contain basically the same components as the hot water systems, except the steam system does not require a circulating pump. This is due to the fact that steam flows naturally from higher pressures to lower pressures. As such, steam circulation is caused by lowering steam pressure along the steam supply mains and in the terminal units. This is accomplished through pipe friction. There are also differences between the boilers in hot water systems and steam systems, because steam boilers require a feedwater system in order to maintain their water levels. Hot water boilers, however, have a boiler feed through the pressure reducing valve in the feedwater line.

BOILERS

A steam boiler is constructed of steel or cast-iron and is partly filled with water, Figure 8-9. The boiler heats this water and converts it into steam for use in heat, or as a power source. The steam boiler is a closed vessel and uses electricity, gas, oil, or electricity as its heat source. In order for the water convert to steam, it must reach its boiling point of 212 °F. To heat the water to this temperature more quickly, it is advantageous to also have a combustion chamber. This chamber makes fuel burn more efficiently by allowing the fuel and air to mix in greater quantities, and higher combustion temperatures result.

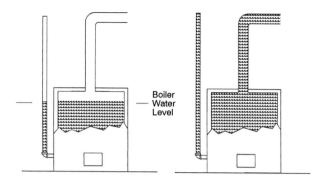

Figure 8-9. Comparison of Water Levels in Steam and Hot Water Boilers

As stated previously, the steam boiler must also have a feedwater system to feed the boiler a constant supply of water while the boiler generates steam. The feedwater system retrieves the water from condensing steam at the terminal unit and returns it to the boiler.

To understand how the feedwater system works, it helps to trace the steam from boiler to terminal unit, Figure 8-10. The steam leaves the boiler through a pipe and travels to the main line (also called a main header). This line is a large diameter pipe where steam accumulates before it is sent on to the terminal unit. From the main line, the steam disperses into branch lines leading to various terminal units. Once the steam reaches the terminal units, it releases heat and condenses back into water (condensate).

In a few systems, a vacuum pump creates a force which helps draw this condensate out of the return lines and into a condensate tank. A trap located near the terminal unit allows only condensate, not steam, to travel back to the condensate tank. The vacuum pump then circulates this water back to the boiler through feedwater lines. The boiler converts the returning condensate into steam again.

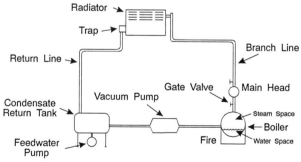

Figure 8-10. Steam Cycle

Boilers are classified into two types: fire-tube and water-tube.

Fire-Tube Boilers. Fire-tube boilers are most often used in heating and producing industrial process steam. This is due to the fact that fire-tube boilers can store a large amount of water, and also these boilers are relatively inexpensive. Fire-tube boilers contain combustion heat and gases confined within tubes. These tubes are then surrounded by water, Figure 8-11.

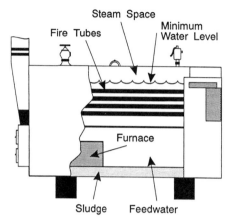

Figure 8-11. Fire-Tube Boiler

There are several types of fire-tube boilers used. These include the commonly found Scotch Marine boiler, vertical fire-tube boiler, locomotive boiler, and horizontal return tubular (HRT) boiler.

Water-Tube Boilers. A water-tube boiler is exactly the opposite of a fire-tube boiler, in that it contains the water within tubes, Figure 8-12. Normally, these water-filled tubes are housed external to the boiler. This configuration has its advantages, as these tubes can be heated rapidly due to their increased heating surface. This helps intensify circulation which helps compensate for sudden fluctuations in steam demand.

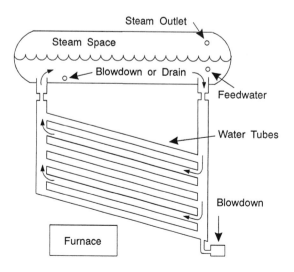

Figure 8-12. Water-Tube Boiler

There are several different water-tube boilers used, including straight-tube boilers, bent-tube boilers, single-drum boilers, and multi-drum boilers.

CHIMNEYS

Chimneys required for gas and oil furnaces are also required for hydronic systems which use gas or oil as their heating source. This is due to the combustion process that takes place in the system. The products of combustion must be vented outside, hence a chimney is required. As in gas and oil systems, some sort of draft is also required to move the products of combustion up the chimney.

In a hydronic system, after combustion takes place, the resulting products rise and enter a large diameter pipe at the top of the boiler called a breeching. In a residential setting, a natural draft lifts these products up into the chimney, where they escape into the atmosphere. For commercial or industrial applications, a fan is needed to blow these products out of the breeching, through exhaust vents, and into the chimney as shown in Figure 8-13.

The principles used for building chimneys in gas and oil heating systems also apply to chimneys in hydronic systems. Chapter 3 contains all the particulars concerning proper chimney construction.

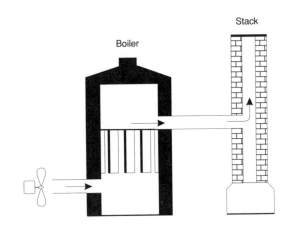

Figure 8-13. *Forced and induced draft system.*

WATER TREATMENT

All water contains scale-forming and corrosive compounds. Calcium and magnesium are scale-forming compounds, meaning they build up in the boiler (both steam and hot water). To remove these compounds, hand-scraping or power-driven wire brushes must be used. The thicker the deposits, the more costly and hazardous to remove them. Well-water normally contains more scale-forming compounds, while surface water contains corrosive compounds. In order to prolong the life of the boiler and piping, the water can either be treated before it enters the system (external), or while it is in the system (internal).

External treatment entails removing scale-forming compounds and dissolved gases (oxygen and carbon dioxide) from the water before it enters the system. External treatment also controls the alkalinity (the ability of water to neutralize acid) of the water. Alkalinity is an important factor, because as alkalinity increases, the more likely scale is to form. Testing for alkalinity is the most accurate indicator of whether or not water will be corrosive or scale-forming.

Internal treatment involves placing chemicals directly into the boiler water. The chemicals used in internal treatment include alkaline phosphate and tannin, an organic material. Tannin is used to disperse the sludge that can form in the boiler. A combination of alkaline phosphate and tannin is normally used to combat scale on the internal boiler surfaces and prevent corrosion of boiler tubes. For the best results, a combination of both external and internal treatments is normally used.

There are test kits available to test for many situations, including alkalinity, tannin, phosphate, and water hardness. The manufacturer provides the correct procedure to follow with each test kit. It is important to follow these procedures in order to ensure accurate readings.

REVIEW QUESTIONS

1. What are the four main components in a hydronic heating system?
2. What does a boiler do?
3. Where are the air vents located, and what is their purpose?
4. What happens when the pressure relief valve opens?
5. What is an aquastat?
6. How is the water level maintained in a boiler?
7. What is the purpose of an expansion tank?
8. What is an impeller? How does it relate to the centrifugal pump?
9. Describe a one-pipe system.
10. How does a two-pipe system differ from a one-pipe system?
11. What are the two types of two-pipe systems? How do they differ from each other?
12. What is a terminal unit? Name three different examples of terminal units.
13. Why is a circulating pump not required in a steam system?
14. What is a feedwater system? How does it work?
15. What are the two different types of steam boilers? What is the difference between the two?
16. What are the two different fans used to expel combustion air from the boiler?
17. Why is water treatment important in a hydronic heating system?
18. What is scale? How can it be prevented?
19. What is the difference between internal and external water treatments?

Chapter **9**
Cooling Fundamentals

The previous chapters covered the process of adding heat to a space, in order to make up the heat loss created by the temperature difference between the inside space and the outside air. The chapters on cooling, however, are concerned with the reverse process: the removal of heat from a space in order to reduce the temperature to a comfortable level.

AIR CONDITIONING

Most people primarily associate air conditioning with cooling. Complete air conditioning goes beyond simply maintaining a comfortable temperature. Air conditioning filters the air, provides optimal air movement and circulation, and maintains a comfortable humidity (moisture) level. An air conditioning system should accomplish these actions, plus temperature maintenance, in both the heating and cooling modes.

FILTERING

Air filtering is generally the same for both summer and winter air conditioning. Air filtering equipment typically consists of very fine, porous substances through which air is drawn to remove contaminating particles. Filters using activated carbon and electrostatic precipitators may be added to the usual filtering mechanisms to further purify the air. For more information on air pollutants and the methods used to remove them, see Chapter 13.

AIR CIRCULATION

Proper air distribution is also necessary to provide a comfortable environment. As in heating systems, the air in a cooling system must be circulated throughout in order to cool the living space. Typically, both heating and cooling modes use the same air mover. Since the air must be clean in both modes, filters are equally important during both cycles.

RELATIVE HUMIDITY

Air conditioning must also control the relative humidity in the living space. Relative humidity is the amount of moisture in the air, expressed as a percentage of the maximum amount of moisture that the air is capable of holding at that temperature. Humidity control for the cooling season is different than it is during the heating season, due to the fact that warm air holds considerably more moisture than cool air.

Maintaining a comfortable humidity level in the cooling mode requires reducing the moisture level, whereas in heating, it involves increasing the moisture level. In winter, the cold outside supply air does not contain a great deal of moisture. When this air is heated, the relative humidity decreases even further, so moisture must be added. The opposite is true in the summer, when the outside air already contains a great deal of moisture. When the temperature of warm outside supply air is reduced (in order to cool the space), the relative humidity increases. Consequently, in the cooling mode, the moisture content of the air must also be reduced to a comfortable level. Most people feel comfortable in a temperature range of 70 to 75 °F, with a relative humidity of 40 to 60 percent.

Humidity control for summer conditions typically includes using automatic dehumidifiers to remove excess moisture. This is usually done at the time the air to be cooled passes over the cold evaporator surfaces. For more information on humidity, see Chapter 12.

A complete air conditioning system involving all these factors should provide automatic control of indoor climate conditions in both summer and winter. For winter heating, the desired room temperature is maintained by automatic control of the heating system. Summer cooling involves keeping the room temperature at the desired level by automatic control of the refrigerating system.

PRINCIPLES OF COOLING

The process of cooling is actually the removal of heat. The refrigeration process is based on the fact that when heat is removed from an object or space, the temperature of that object or space decreases. To accomplish space cooling, air conditioning systems utilize the refrigeration process to remove excess heat energy from the space. Stated in industry jargon, the heat is removed from the heat source and disposed of in the heat sink.

SENSIBLE HEAT

As mentioned previously, heat is defined as energy that moves from a warmer body to a colder body. Adding heat energy to a substance changes the kinetic energy of that substance. While heat energy has neither weight nor dimension, its presence can be detected and measured with thermometers. U.S. measurements of heat are expressed in terms of degrees Fahrenheit, a scale on which (at sea level) water freezes at 32 °F and boils at 212 °F.

Thermometers are used to measure temperature. The points on a thermometer are indications of sensible heat, because any addition or subtraction of heat can be sensed and measured directly, and can also be sensed by touch. Thermometers measure heat intensity, while heat quantity is measured in Btu.

As defined earlier, a Btu is the amount of heat necessary to raise the temperature of 1 lb of water 1 °F. Consequently, a cooling Btu is defined as the amount of heat which must be removed in order to reduce the temperature. In other words, whether heating or cooling, it takes one Btu to either raise or lower the sensible heat (temperature) of 1 lb of water 1 °F. Heating raises the temperature and cooling reduces the temperature, but in either case the heat intensity is measured in degrees and the heat quantity is measured in Btu.

LATENT HEAT

Latent heat is the amount of heat that can be added or subtracted from a substance without changing its temperature. This is also described as increasing or reducing the kinetic energy of the substance. For example, it is necessary to remove the latent heat in a 32 °F pan of water before it can become 32 °F ice. In this case, 144 Btu are removed, but the temperature remains the same. Similarly, 212 °F water can exist either as a liquid or a vapor, depending upon its total heat content (both latent and sensible heat). Latent heat is sometimes called hidden heat, because it cannot be measured directly with a thermometer.

Latent heat must be considered in heating and cooling, because its presence affects whether or not a substance will undergo a change of state. This is significant because change of state is vital to the refrigeration process. In order to accomplish cooling, the refrigerant in the system must change state from a liquid to a vapor (evaporation) and then back again, from a vapor to a liquid (condensation). Figure 9-1 illustrates the amount of total heat required to change water from a solid to a liquid to a vapor.

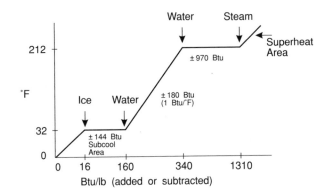

Figure 9-1. *Btu/lb*

PRESSURE

Pressure is defined as force per unit of area and is most commonly measured in pounds per square inch (psi). Pressure is another factor which influences change of state. To demonstrate, at atmospheric pressure, water (in an open container) boils at a temperature of 212 °F. However, if the water is confined under pressure, it will not boil away—the temperature can be raised above the boiling point without most of the water changing into steam. Pressure cookers are an everyday example of this concept.

REFRIGERANTS

As previously noted, the processes of evaporation and condensation are essential to the operation of a mechanical refrigeration system. Refrigerants are special liquids developed to help the cooling system accomplish these processes more efficiently.

An efficient air conditioning system must have the capability of readily absorbing and rejecting large amounts of heat at the normal operating temperatures of the equipment. The system must continuously repeat the processes of evaporation and condensation with the same substance. While water could theoretically be used for this purpose, it boils at a temperature too high for cooling and freezes at a temperature too high to be useful in the low temperature conditions. This is why refrigerants are used.

Refrigerants must possess certain characteristics to be effective in a cooling system. They must be:

- small in volume but high in density.
- able to operate at low differences in pressure.
- nonflammable and nonexplosive in either vapor or liquid states.
- noncorrosive and nontoxic.
- able to carry oil in solution.
- highly resistant to electricity.

Many of the refrigerants used earlier in air conditioning and refrigeration lacked the qualities that are required in today's demanding field. Some of the refrigerants used were ammonia, sulfur dioxide, methyl chloride, propane, and methane. These liquids contained some undesirable qualities, and they have been almost completely replaced by new, specialized refrigerants.

Sometimes two or more existing refrigerants are combined to form what is known as an azeotrope. An azeotrope acts as though it were a pure compound. When the different refrigerants are mixed together, the resulting compound contains different characteristics than either of the refrigerants individually. Azeotropes are used in situations requiring more refrigeration ability than one refrigerant alone can provide.

REFRIGERANT 12 (R-12)

Dichlorodifluoromethane (R-12), is a clear, almost colorless liquid and gas and is used in commercial and residential applications. Its boiling point of -21.62 °F at atmospheric pressure means that at room temperature, R-12 exists as a vapor. R-12 is nontoxic, nonirritating, nonexplosive, and for all practical purposes, nonflammable. Only a moderate volume of R-12 per ton of refrigeration is required. Some of the available brands of R12 include Freon 12, Genetron 12 and Isotron 12.

REFRIGERANT 22 (R-22)

Monochlorodifluoromethane (R-22) is also a clear and colorless liquid and gas and is primarily used in commercial and industrial applications. Its boiling point is -41.36 °F, and it is nontoxic, nonirritating, nonexplosive and nonflammable. R-22 has higher saturation pressures than R-12, and only a small volume of R-22 per ton of refrigeration is required. Because of its high thermal stability, R-22 is suitable for use in air-cooled condenser systems. Freon 22, Genetron 22 and Isotron 22 are some of the available brands of R-22.

REFRIGERANT 500 (R-500)

This particular refrigerant is an azeotrope, consisting of 73.2 percent R-12 and 26.2 percent R-152a (another refrigerant). R-500 has a boiling point of -28 °F at atmospheric pressure. It is used primarily in commercial and industrial settings and offers greater refrigeration ability than R-12 alone.

REFRIGERANT 502 (R-502)

This is another azeotrope, composed of 48.8 percent R-22 and 51.2 percent R-115. R-502 has a boiling point of -49.76 °F at atmospheric pressure. It is nonflammable and nontoxic and a good use for R-502 is a refrigerated display case, as it works well in low-temperature applications.

REFRIGERANT 717 (AMMONIA)

As mentioned earlier, ammonia has been around for many years. It is only used for industrial applications. Ammonia is a colorless gas with a strong odor. It is flammable, explosive, and highly toxic. Ammonia's boiling temperature, at atmospheric pressure, is -28 °F.

NOTE: All refrigerants are supplied in color-coded drums. Each type of refrigerant is identified by a certain drum color, regardless of the manufacturer. For instance, R-12 is always stored in a white container, and R-22 always comes in a light green container, no matter which manufacturer supplies them.

REFRIGERANT RECOVERY

The future of refrigerants is uncertain, as they have been shown to directly contribute to the depletion of the earth's ozone layer. For this reason, it is now necessary to recover and reclaim refrigerant instead of venting it into the atmosphere. Certified recovering equipment must be used in order to recover the refrigerant, Figure 9-2. The service technician must then take this "used" refrigerant to a reclamation center where the contaminants are removed, and the refrigerant is returned, once again, to industry specifications.

New, ozone-safe refrigerants are being developed to take the place of the standard refrigerants. This is due to the fact that ozone-damaging chlorofluorocarbon (CFC) refrigerants are to be phased out eventually. These CFCs include R-11, R-12, R-113, R-114 and R-115. Hydrochlorofluorocarbon (HCFC) refrigerants are also deemed unsafe for the environment, and they must be phased out by the year 2030. R-22 and R-123 are HCFCs.

Figure 9-2. Recovery Equipment. Courtesy, Carrier Corporation, a Subsidiary of United Technologies Corporation.

The new, ozone-safe refrigerants include a hydrofluorocarbon (HFC) called R-134a. This particular refrigerant is designed to eventually take the place of R-12. HFCs contain no chlorine, so they are safer for the ozone. Another refrigerant, R-123, is being designed to supplant R-11. R-123 is an HCFC, however, its ozone-depleting potential is considered to be much smaller than R-22.

There are several techniques a service technician can use in order to prevent CFCs from escaping. These include:

■ Do not overfill a system and then bleed off the refrigerant until the right pressure readings are obtained. This practice is harmful to the environment and wastes refrigerant. Also, accurate equipment is available which makes charging a system fast and simple.

■ Make sure the system is properly evacuated before charging. (Evacuation is discussed in Chapter 11.) Thermistor vacuum gages are highly accurate in measuring the vacuum and can also be used to check for leaks on empty systems.

■ Use fittings to seal off the hose when the hose is disconnected. This allows the refrigerant to remain within the hose.

■ Use low-permeation hoses. These hoses prevent refrigerant from seeping through the hose material and escaping into the atmosphere.

■ Fix all leaks. A system low in refrigerant can be a sign of a leak. Instead of "topping off" the system, check to make sure a leak is not the culprit.

SAFETY PRECAUTIONS

Though the refrigerants used today, with the exception of ammonia, are much safer than those used previously, there are certain safety precautions that should be taken whenever refrigerants are handled.

■ Because refrigerants are kept under high pressure, their containers should not be heated or left out in the sun. External heat on the container causes the internal pressure to rise, which causes the refrigerant to expand. If the pressure becomes excessive, the container can burst.

■ R-22 boils at a very low temperature, -41 °F, and if released to atmospheric pressure, it turns into a vapor. This vapor may cause severe burns if it comes in contact with any part of the body. It is extremely important that refrigerant be kept out of contact with the eyes. Therefore, safety glasses should be worn whenever work involves refrigerant.

■ Although refrigerant does not ignite, it will form phosgene gas if it comes in contact with an open flame. Phosgene gas is odorless and colorless, but it is very poisonous. Any situation that may result in the formation of this gas should be avoided.

■ Refrigerants are processed to be free of water down to about 10 parts per million (ppm). Water must be removed because when it comes in contact with the heat from the compressor, it breaks the refrigerant down into hydrofluoric and hydrochloric acids. These acids pick up copper from the tubing in the system and deposit the copper on the steel cylinder walls of the compressor. The compressor piston has extremely low tolerances, and even a very small amount of copper plating can seize the compressor. Consequently, precautions should be taken during installation so that no moisture is introduced. All systems should include a drier to remove any moisture that might be in the system, or not removed during evacuation.

SATURATED VAPOR

Vapor is considered to be saturated whenever it is in contact with the liquid from which it evaporated. The saturation point occurs when the vapor is at its boiling point temperature, which corresponds to its pressure. For example, in a closed container, such as a half-filled refrigerant drum, the space above the liquid refrigerant is filled with saturated vapor. This remains saturated vapor as long as the temperature and pressure of the vapor and refrigerant remain the same. If more heat is added to the vapor, its temperature will rise. However, if heat is taken away from the vapor, it will revert to liquid form. In a saturated condition, vapor is constantly condensing into a liquid, to be replaced by liquid that is evaporating. Consequently, there is an equilibrium in the percent of vapor at a given pressure. There can be two other conditions of this liquid-vapor mixture: superheated vapor and subcooled liquid, Table 9-1.

R-22		WATER
Temp °F	Pressure psig	Boils at: psig
300		52 LBS
250		15
212		ZERO
175		16" Vacuum
150	382 LBS.	22"
130	297	25"
120	260	26"
110	226	27"
100	196	28"
90	168	28.5"
80	144	29"
70	121	29.2"
60	102	29.4"
50	84	29.6"
45	76	29.7"
40	69	29.8"
35	62	
30	55	
25	49	
20	43	
15	38	
10	33	
5	28	
ZERO	24	
-20	10	
-40	ZERO	

Table 9-1. Saturated Pressure/Temperature Chart

Any time the vapor and liquid are in contact with each other, they are both at the same temperature. A liquid-vapor mixture is always saturated and can never become superheated or subcooled until after they are separated.

SUPERHEATED VAPOR

Superheated vapor is vapor which has been heated in excess of its boiling temperature. When supplied with sufficient heat, all the liquid refrigerant may be evaporated. When this occurs, the refrigerant is in a state of 100 percent saturation, indicating that no more liquid is available. If additional sensible heat is applied, the vapor becomes superheated, and the temperature of the vapor

exceeds the saturated temperature at the same pressure. This condition can also be used to set the metering valve, to protect the compressor.

SUBCOOLED LIQUID

When a liquid refrigerant cools, all of the vapor condenses; it is now at zero saturation, indicating that no more vapor is available. If the liquid cools below this point, the liquid is said to be subcooled. At this point, the temperature is below the saturated temperature at the same pressure. Normally, the liquid line is subcooled for some portion of its length as it leaves the condenser.

ENTHALPY

The total heat content, including both latent and sensible heat, is called enthalpy and is expressed in Btu/lb. Enthalpy is a method of measuring the heat absorbed from the living space by the refrigerant, as well as the heat rejected to the outside atmosphere. Since the enthalpy values are already calculated, they are usually read from a pressure-enthalpy diagram for the specific refrigerant.

REFRIGERATION CYCLE

The compressor, condenser, metering device, and evaporator are the main components of a refrigeration unit. These components, which will be discussed in later chapters, actually control the refrigeration cycle. At this point, however, it is necessary to determine temperature, pressure, enthalpy, and state in the system once the refrigeration cycle starts.

When the thermostat calls for cooling, the following steps occur simultaneously:

- The compressor starts pumping vaporized refrigerant out of the evaporator. It compresses the vapor and then sends it to the condenser. This creates a pressure difference between the high and low sides.
- The condenser fan turns on and blows outside air across the coil of the condenser. As a result, heat in the refrigerant vapor dissipates to the outside air, causing the refrigerant to change state from a vapor to a liquid. This liquid then travels to the metering device.
- The metering device passes liquid refrigerant into the evaporator so it can start to collect the heat from the air stream around the evaporator.
- The evaporator blower, if not on continuous blower operation, comes on at this point and passes the warm air from the space across the face of the evaporator. This allows the refrigerant to absorb the heat in the air.

Theoretically, when the liquid refrigerant reaches the metering device, the temperature of the refrigerant is 114 °F, approximately 299 psig, and an enthalpy of 45 Btu/lb. The liquid refrigerant moves through the metering device into the low side of the system. Here, the refrigerant expands and cools and travels through the evaporator coil, where it picks up heat from the living space. As the liquid picks up heat, it changes to vapor, and the vapor exits the evaporator at a temperature of about 45 °F, and a pressure of 77 psig. Before this vapor enters the compressor, it is superheated to 55 °F, but the pressure remains at 77 psig. Its enthalpy now increases to 100 Btu/lb, having picked up 64 Btu of latent heat from the living space air and 1 Btu of sensible heat from the superheating process.

The vapor moves into the compressor where it picks up additional heat from the compressor motor, amounting to approximately 24 Btu. Now, the vapor has a heat content of 89 Btu. The compressor compresses this vapor, which raises the temperature of the vapor to 230 °F, and the pressure also rises considerably. Its enthalpy now is 134 Btu/lb.

The compressor expels this vapor through the discharge line and into the condenser. The temperature of the vapor is now 130 °F, and its pressure is about 299 psig. As the vapor passes through the condenser, it loses heat to the outside air and changes state from a vapor back to a liquid. By the time the refrigerant leaves the condenser, it is in liquid form, with a temperature of 130 °F and a pressure of 299 psig. When the refrigerant changed from vapor to liquid, it lost 83 Btu to the outside air in latent heat.

Upon leaving the condenser, the liquid is subcooled, eliminating another 6 Btu. As the liquid approaches the metering device again, its temperature is back to the starting point of 114 °F, with a pressure of 299 psig, and an enthalpy of 45 Btu/lb. The entire refrigeration cycle is shown in Figure 9-3.

The values in the refrigeration cycle vary from system to system, depending on the efficiency of the system. Also, outdoor and indoor air temperatures affect the operating temperatures and pressure in a system. The values used in this refrigeration cycle are common for a system operating under ideal conditions.

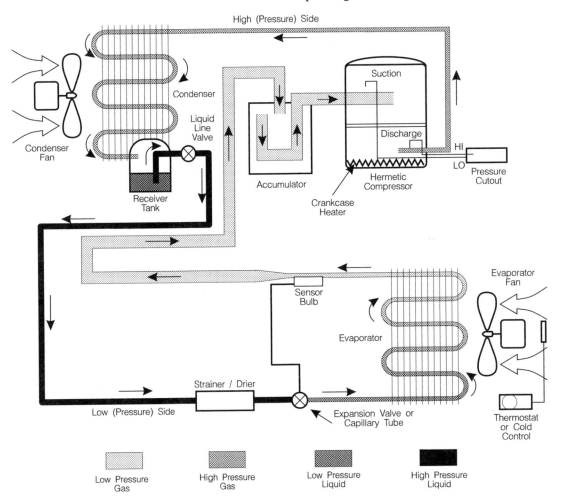

Figure 9-3. Basic Refrigeration Cycle

Review Questions

1. Define air conditioning. – COOLING
2. Define latent heat. CHANGE OF STATES
3. How much heat is removed to change water from a liquid to a solid? 144 BTU PER lb.
4. How much heat is required to change water from a liquid to a vapor? 970 BTU PER lb.
5. What does the term enthalpy mean? SENSIBLE + LATENT HEAT
6. What is saturation point? NO LIQUID LEFT
7. What is superheated vapor? ABOVE BOILING PT
8. What is subcooled liquid? BELOW BOILING
9. Name the basic characteristic of a refrigerant. LOW BOILING PT.
10. Is R-22 a better product than R-12 for central air conditioning? Explain your answer. YES. BETTER THERMAL STABILITY
11. What are the four basic safety factors in working with refrigerants? 1. DONT OVERHEAT CONTAINER. 2. GOGGLES 3. NO FLAMES 4. NO WATER
12. Explain what happens to the refrigerant as it passes through the evaporator. BOILS
13. What happens to the refrigerant as it goes through the compressor? PRESSURE + TEMP. ↑
14. What happens to the refrigerant as it passes through the condenser? LIQUIFIES
15. What happens to the refrigerant as it goes through the metering device? SPRAYS
16. What four functions occur when a system calls for cooling? COMPRESSOR PUMPS, CONDENSER FAN BLOWS, TXV OPENS, EVAP. FAN BLOWS
17. What is the normal operating pressure and temperature at the entrance to the metering device?
18. What are the normal temperatures and pressures at the exit of the evaporator?
19. What is the state of the refrigerant at the exit of the condenser? LIQUID.
20. What is the state of the refrigerant at the exit of the evaporator? VAPOR

Chapter 10

Cooling System Components

COMPRESSORS

The compressor is often referred to as the heart of the refrigeration system, because it pumps refrigerant throughout the system. The compressor is in charge of supplying the forces necessary to keep the system operating. There are various kinds of compressors on the market, yet all have the same basic characteristics.

There are two different pressure conditions, low side and high side, in any compression refrigeration system. It is necessary to distinguish between these conditions in order to understand how the compressor works. Heat is absorbed in the low side, so the evaporator, accumulator, suction line, and entrance to the compressor suction valve are on the low side. Heat is released on the high side, so the condenser is located here. The compressor receives refrigerant vapor at a low temperature from the evaporator. The compressor then compresses this vapor, thus raising its pressure and temperature. This high-pressure, hot vapor then moves to the condenser, where it is cooled and condensed to form a liquid.

COMPRESSOR TYPES

There are three major types of compressors—reciprocating, rotary, and centrifugal. The information included in this section is based on the reciprocating compressor, as they are most commonly used in residential air conditioning.

Reciprocating Compressor. These compressors are quite similar to an automobile engine, in that they contain a piston which travels back and forth in a cylinder. These compressors are used when smaller horsepower sizes are needed, such as for commercial and domestic refrigeration.

Rotary Compressor. This type of compressor has a vane that rotates within a cylinder. Applications for this compressor include home refrigerators and freezers.

Centrifugal Compressor. A high-speed centrifugal impeller with many blades rotating within a housing best describes a centrifugal compressor. It is only used in very large applications.

COMPRESSOR MAINTENANCE

The compressor range for residential applications is from 1 to 5 tons of refrigeration. These compressors are hermetically sealed in a steel shell and therefore cannot be serviced in the field. If there is an internal malfunction, the compressor is removed, and, if still under warranty, it is sent back to the manufacturer. Figure 10-1 illustrates the internal parts of a typical hermetically-sealed compressor.

It should be noted that a high percentage of compressors returned under warranty have no internal malfunction. The majority of such apparent failures is usually the result of incorrect installation. The original compressor installation, if done properly, should be trouble-free and mechanically sound for many years. It is important for a service technician to understand the internal workings of a compressor in order to recognize problems.

COMPRESSOR LUBRICATION

It is very important to maintain a proper oil level in the compressor, as an oil level that is too high or too low can be a hazard. Oil is used to reduce friction caused by moving parts, and to remove some of the heat produced by this friction. It is primarily the responsibility of the compressor manufacturer to recommend which oils are to be used in a refrigeration system.

The quality of oil used is extremely important, and these oils must:

- not carbonize with hot surface contact or deposit wax at lower temperatures.
- not contain any moisture or sulfur compounds.

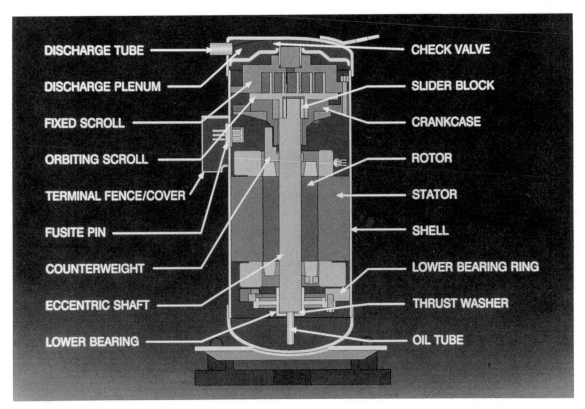

Figure 10-1. Internal Components of a Hermetically-Sealed Compressor. Courtesy, Carrier Corporation, a Subsidiary of United Technologies Corporation.

- be stable in the presence of oxygen.
- be fluid at all times, even at low operating temperatures.
- maintain sufficient body to lubricate at high temperatures.
- not contain corrosive acid.

Special refrigeration grade oils that are dewaxed and dehydrated to correspond with compressor and refrigerant standards are the only oils approved for modern compressors. These special oils are available from refrigeration wholesalers. Only small amounts of oil should be purchased at any one time because once opened, the oil absorbs moisture which can contaminate the system. To prevent this, oils can be obtained in aerosol cans which keep them sealed off from the atmosphere. These are handy when only small amounts are needed.

There are two basic methods that can be used when lubricating the compressor: forced and splash. Most modern refrigeration compressors use forced lubrication, which allows more accurate control of oil distribution. To force lubricate a smaller compressor (3 hp), oil is forced to the necessary points by way of provided passageways. When splash lubrication is used, the crankshaft splashes oil onto the cylinders and pistons, which then move the oil through the compressor valves. Splash lubrication is inadequate for modern, high-speed compressors, due to their higher bearing and friction surface temperatures.

Forced lubrication occurs as described in the following paragraph. A hole is located in the bottom of the crankshaft to carry the oil up to the connecting rod where it exits. A groove is milled in the center of the connecting rod that carries oil to the center of the rod shaft, and then out to the wrist pin. At this point, oil is carried to the compressor cylinder, lubricating that area. Oil from the center of the crankshaft also exits at the top main bearing and is carried by a slot that lubricates the entire bearing surface. The oil going to the bearings will drip down and return to the oil reservoir at the bottom. The oil traveling to the pistons and cylinder must travel through the entire system before returning to the crankcase. The lines must be sized and bent so as not to restrict the flow of oil returning to the compressor crankcase.

Compressors in the 3 to 5 ton range take 45 to 55 ounces of oil charge. When the compressor is not running, the oil level is near the bottom of the lower cylinder, Figure 10-2.

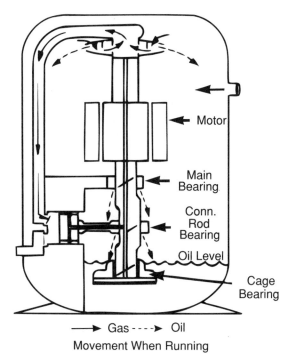

Gas ⟶ Oil ⟶
Movement When Running

Figure 10-2. Oil and Gas Flow Paths

REFRIGERANT GAS FLOW

Gas entering the compressor is called suction gas. It enters the compressor shell at the top, through a suction tube. This tube opens directly into the compressor shell. As a result, the compressor is subjected to the low side pressures of the system.

Suction gas completely fills the inside of the shell and removes some of the heat of the motor. Then the gas is drawn into the cylinder through the oil separator. The oil separator, also known as the antislugging device, takes in gas and oil mixtures through the slots on top. The separator then rotates, using centrifugal motion to force the heavier oil out of the edge slots. In this manner, the majority of the oil is returned to the crankcase rather than going through the cylinder.

Gas exiting the separator follows the tube down into the muffler and enters the cylinder through the suction valve. This compressed gas, called discharge gas, exits through the discharge valve and goes into the discharge tube, which is designed for high pressures. The discharge tube is coiled once to create a vibration loop, then connected to the shell at the external discharge line.

The majority of compressors take in suction gas near the top third of the shell and discharge it near the lower third. Externally, the compressor has one discharge tube and one suction tube. Compressors are mounted on rubber grommets on their external feet, and have a terminal block for electrical connections.

COMPRESSOR FLOOD-BACK

Flood-back occurs when liquid refrigerant enters the compressor through the suction tube, falls to the bottom of the compressor and mixes with the oil. When oil is diluted in this manner, or to any great degree, its lubricating ability is reduced and as a result the bearings overheat. When the heat becomes excessive, the refrigerant and oil break down into hydrochloric acid, which corrodes the motor windings and can lead to compressor burnout.

The following situations can cause liquid flood-back:

- Over- or undercharging of refrigerant
- Restriction of the evaporator air
- Low indoor air temperature or relative humidity reducing the load
- Defective expansion valves
- Very low outdoor temperatures

When these conditions occur, the problem can usually be solved by installing a suction line accumulator, Figure 10-3. As the term implies, an accumulator collects fluid, preventing liquid refrigerant from flowing into the compressor (or any other component) before it can cause damage. The accumulator then provides a place for this liquid to evaporate to a gas, before sending it on to the compressor. The accumulator is a small tank inserted into the suction line just before the compressor. It acts as a trap for the liquid refrigerant, dropping the liquid to the bottom of the tank. At the same time, it allows the gas to be drawn into the compressor through an outlet tube. A small hole at the bottom of the tube permits oil to be picked up by the refrigerant vapor as it passes through, allowing it to be returned to the compressor crankcase.

Figure 10-3. Suction Line Accumulators. Courtesy, Carrier Corporation, a Subsidiary of United Technologies Corporation.

REFRIGERANT MIGRATION

When a compressor is turned off, the refrigerant within it condenses into a liquid. This liquid then migrates to the coldest part of the system, the crankcase, and mixes with the oil.

The refrigerant collecting in the crankcase subsequently condenses into a liquid in the compressor shell. Eventually, the shell fills with liquid. When the compressor turns on, the liquid refrigerant-oil mixture begins to boil. The refrigerant boils off quickly, which then causes the oil to foam. This foaming action can cause the mixture to enter the cylinder and cause slugging, which can result in blown gaskets. This can occur even if an antislugging device has been installed. Whether slugging occurs or not, refrigerant migration causes quite a bit of oil to permeate the entire system. The compressor is then operating with an insufficient oil supply which can result in damage to the compressor bearing surfaces.

To prevent refrigerant migration, heat can be added to the crankcase during the time the compressor is not in use. A crankcase heater, an oil rectifier, or run capacitors can help accomplish this.

Crankcase Heater. This is an electric heater element of about 50 W. This element wraps around the base of the compressor and is wired so that it always contains electrical current, even when the compressor is off. This heater can hold the inside temperature of the crankcase at about 10 °F above the ambient, which is enough to prevent the refrigerant from condensing, Figure 10-4. A similar method is a blanket-type heater. This is also a resistance heater, but has a greater surface area. It is placed beneath the compressor crankcase rather than wrapped around it.

Figure 10-4. Crankcase Heater. Courtesy, Carrier Corporation, a Subsidiary of United Technologies Corporation.

Oil Rectifier. This is an electric resistance heater in the form of a rod element. It is inserted into the compressor crankcase near the lower third of the normal oil level. Compressors contain a well which accepts the crankcase heater. The rod is inserted into this well and wired directly to the line side of the contactor so that it is operative whenever the main disconnect switch is on.

Run Capacitors. During the off cycle, run capacitors can be wired so that they impose a low voltage on the compressor motor windings. A run capacitor, which has a known and constant impedance, is located in one leg of the line. This causes a very large voltage drop across the capacitor terminals and a small voltage drop across the compressor terminals, allowing a small current to flow through the motor windings. Since the motor is immobile, this current is dissipated as heat.

Capacitors are matched to the motors so that due to the electrical input to the winding, they create the required temperature differential between the compressor and the surrounding ambient. This temperature difference maintains a higher crankcase temperature and prevents any sizable quantity of refrigerant from migrating to the compressor housing.

CONDENSERS

The purpose of the condenser is to cool the compressed refrigerant vapor (received from the compressor), change this vapor to liquid, then subcool this liquid. Basically, the role of the condenser is to remove heat from the refrigeration system. There are three types of condensers: air-cooled, water-cooled and evaporative.

AIR-COOLED CONDENSERS

These condensers are used extensively, most often in small commercial and domestic applications. Air-cooled condensers consist of copper tubes which accomplish primary heat transfer, Figure 10-5. Aluminum fins, acting as secondary heat transfer surfaces, attach to the copper tubes. These fins help to increase heat dissipation. It is common for several rows of tubes and fins to be used. These tubes are usually staggered to maximize air turbulence and heat transfer from the passing air, Figure 10-6. The outside air forced over the surface with a condenser fan moves from 1,000 to 2,000 cubic feet per minute (cfm) over the fins.

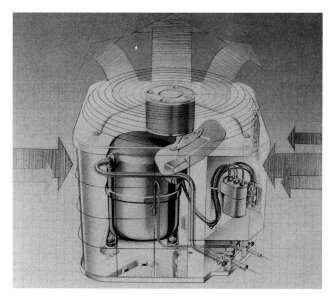

Figure 10-5. Condenser Air Movement. Courtesy, Carrier Corporation, a Subsidiary of United Technologies Corporation.

WATER-COOLED CONDENSERS

Water-cooled condensers are a good choice for industrial applications, particularly when ammonia is used as the refrigerant. These condensers operate in much the same manner as air-cooled condensers, except water, rather than air, is the medium used. Water-cooled condensers have several advantages over air-cooled condensers, in that water causes better heat transfer. Also, water is usually much cooler than the normal ambient air temperature. Water-cooled condensers do have drawbacks, as they require a drain, a water-regulating valve, and they can also cause corrosion.

EVAPORATIVE CONDENSERS

The evaporative condenser is a combination of the air-cooled and water-cooled condensers. These condensers are often used in hot, dry regions.

EVAPORATORS

The main function of an evaporator is to absorb heat into the refrigeration system, in order that the heat may be moved to the condenser. The evaporator absorbs this heat from the surrounding air or liquid, then the refrigerant helps remove the heat from the system. Condensate forms during this process, and the evaporator must provide a place for this condensate to collect.

The two basic types of evaporators are the dry or direct-expansion evaporator and the flooded evaporator. The evaporator refrigeration coil also comes in several different configurations, depending on the application. These designs include the A coil, slant coil and H coil.

DRY OR DIRECT-EXPANSION EVAPORATOR

This type of evaporator is the mostly commonly used and consists of a continuous tube, Figure 10-7. The refrigerant is fed into one end of the tube, and the suction line connects to the other end of the tube. The refrigerant flows through the tube, vaporizing and absorbing heat. In this type of evaporator, the liquid refrigerant is not recirculated, because the refrigerant fed into the tube is just sufficient to totally evaporate before leaving the evaporator.

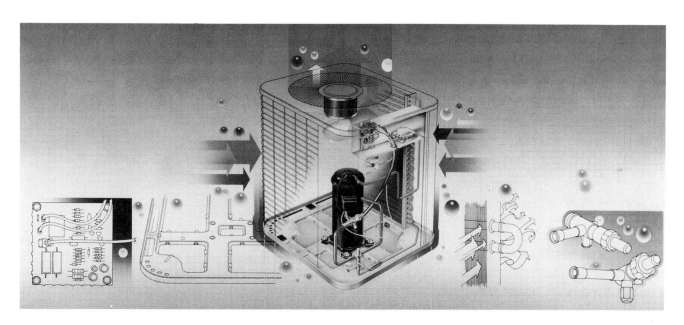

Figure 10-6. Basic Condensing Unit Components. Courtesy, Carrier Corporation, a Subsidiary of United Technologies Corporation.

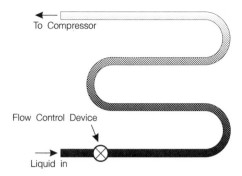

Figure 10-7. Direct-Expansion Evaporator

FLOODED EVAPORATOR

In this type of evaporator, there is an abundance of liquid refrigerant circulating in the tube, so there is always liquid present, Figure 10-8. The liquid refrigerant flows through a flow control device into a storage tank, called an accumulator or surge drum. From there, the liquid refrigerant flows through the tubing, vaporizing and absorbing heat. The liquid refrigerant that does not evaporate moves back to the accumulator, but the vapor separates from the liquid in the upper part of the accumulator and is drawn to the suction line.

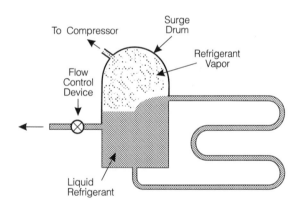

Figure 10-8. Flooded Evaporator

EVAPORATOR CONFIGURATIONS

The evaporator is actually a refrigeration coil, made of copper or aluminum, with fins attached to the coil in order to provide better heat exchange. Depending on the installation, the coil can take several different shapes to ensure airflow through the coil and drainage of condensate from the coil. These designs include: A coil, slant coil, and H coil.

A Coil. This coil contains two evaporators which are connected together and mounted at an angle to each other, roughly forming the letter A, Figure 10-9. An A coil can be used for upflow, downflow, and horizontal flow

applications. The condensate collects at the bottom of the A configuration when used in an upflow or downflow application. In a horizontal flow application, the coil is turned on its side, and the condensate collects at the bottom of the coil. An A coil is not necessarily desirable for a horizontal flow application.

Figure 10-9. A Coil. Courtesy, Carrier Corporation, a Subsidiary of United Technologies Corporation.

Slant Coil. Upflow, downflow, and horizontal flow applications can all use a slant coil, Figure 10-10. This coil is designed to provide more surface area and is mounted on an angle. The condensate collects at the bottom of the slant.

Figure 10-10. Slant Coil. Courtesy, WeatherKing Division of Addison Products Company.

Slab Coil. The slab coil looks like a flat rectangle, Figure 10-11. It is designed for horizontal flow applications, however, it can be mounted vertically. Condensate normally collects at the bottom of the slab.

Figure 10-11. *Slab Coil. Courtesy, WeatherKing Division of Addison Products Company.*

This tubing configuration causes the refrigerant circuit to be closed and allows the refrigerant to be recycled and reused. In a closed system, the pressure in the evaporator is held to about 77 psig. At 77 psig, the refrigerant will not freeze the coil, Figure 10-12.

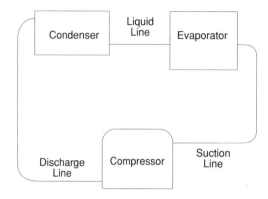

Figure 10-12. *Connecting Lines Used to Link Major System Components*

REFRIGERATION SYSTEM CONFIGURATION

Copper tubes connect the evaporator, compressor and condenser to form a complete, closed, refrigerant circuit. These tubes include the suction line, discharge line, and liquid line. There are other components, such as liquid line driers, pressure controls and metering devices that are also necessary.

SUCTION LINE

The suction line connects the evaporator to the compressor and prevents liquid refrigerant from entering the compressor. Because this line carries gas, which naturally has a greater volume than liquid, it must have a larger diameter. The suction line is insulated to help maintain the suction gas at a lower temperature.

HOT GAS OR DISCHARGE LINE

The hot gas or discharge line runs from the compressor discharge valve to the inlet side of the condenser. This line carries refrigerant vapor and oil droplets. Since the compressor and condenser are mounted in the same cabinet, only a short line is needed. The discharge line is smaller than the suction line, but is much hotter, as it absorbs heat from the compressor motor and the heat of compression.

LIQUID LINE

A third line runs from the outlet of the condenser, usually at its lowest point, back to the inlet side of the evaporator. This line is called the liquid line, because it carries the liquid refrigerant discharged from the condenser.

LIQUID LINE FILTER-DRIERS

While manufacturers take extreme care to prevent any moisture entering a system during its manufacture and installation, most still choose to install a liquid line filter-drier. A filter-drier is designed to remove solids and undesirable solubles, such as water and acid, from the refrigerant.

There are many types of filter-driers, but most combine activated alumina desiccant with a molecular sleeve. The combination of these materials creates a drying agent. A metal screen positioned at the entrance traps solids and dirt particles, and the core design ensures all of the refrigerant comes in contact with the drying agent. Filters are mounted in the liquid line, usually near the condensing unit.

When a compressor must be replaced because of a burn-out, the liquid line drier should be changed also. By the same token, the drier should always be changed if a system has been opened or is even suspected of containing moisture.

PRESSURE CONTROLS

To protect against unsafe pressure conditions, compressors include a low-pressure control, and some also have a high-pressure control, usually mounted in the compressor shell. A small tube, or capillary, connects to a port on the service valve so it can sense pressure under operating conditions. Both controls may be combined into one control called a dual-pressure control. They are wired electrically in series into the low voltage control circuit.

The low-pressure control protects against loss of refrigerant charge, should the suction pressure drop below the control setting. The high-pressure control protects against failure of the condenser fan motor, or any other condition, which causes the head pressure to increase to an unsafe point. It usually has a manual reset feature, meaning that once it trips, it must be reset manually before the compressor resumes operation.

METERING DEVICES

The metering device, also called a flow-control device, is the dividing point between the high-pressure and low-pressure sides of the system. Its job is to control the flow of liquid refrigerant to the evaporator. This control of liquid refrigerant is important, because proper operation of the evaporator depends on the correct amount of liquid refrigerant. Also, the evaporator must evaporate all liquid refrigerant. If this does not occur, compressor slugging can result. There are different types of metering devices, which will be discussed shortly.

When illustrating a metering device, it is easiest to think of a compressor, which has a fixed cylinder capacity plus a constant 3,500 rpm operating speed. While the compressor is fixed, the refrigerant is in a vapor state when it enters the compressor, and as such, temperature can affect its volume and density (i.e., when temperature increases, volume increases). If, for example, the volume of the refrigerant increases, the pumping rate of the compressor decreases, due to a decrease in density. The temperature of the refrigerant gas at the compressor is proportional to the amount of refrigerant the metering device passes to the evaporator. The more refrigerant present, the higher its temperature. Therefore, it is very important for the metering device to pass along only the amount of refrigerant needed, based on how the entire refrigeration system is working.

Each time a pound of refrigerant makes the circuit through the refrigeration system, the system tries to remove as much heat as possible. There are two basic factors that affect the ability of the system to remove heat, and thus increase or decrease the capacity and flow rate of the refrigerant. These factors are changes in the temperature or quantity of outdoor air, and changes in the temperature or quantity of indoor air.

The quantity of outdoor air is fixed, as the condenser fan runs at a constant speed; as such, it is only necessary to consider changes in outside air temperature. When outside air temperature increases, it accepts less heat from the refrigerant because the temperature difference between the refrigerant and outside air is smaller. As a result, the compressor must pump the gas to a higher temperature, or head pressure, to move it into the condenser. This reduces the quantity of refrigerant the compressor can pump. In response, the metering device must allow less refrigerant to flow in order to equal the lower pumping rate. This situation reduces the overall capacity of the system.

On the other hand, when the outside air becomes colder, the reverse happens. The metering device must allow more refrigerant to flow, because the compressor is able to pump more refrigerant, due to the condenser's improved ability to reject heat. When this situation occurs, the capacity of the system is increased.

The quantity of indoor air passed into the evaporator also remains constant, due to the constant speed of the blower. As the indoor air becomes warmer and lighter in weight, the same blower speed can pass along a greater volume of air. When the air has more heat content, it is easier for the evaporator to extract this heat from the air, so the metering device must allow a greater flow of refrigerant into the evaporator. However, if the indoor air becomes colder, it also becomes heavier, so the quantity or volume of air passed over the evaporator becomes slightly less. Less heat can be extracted from colder air because the temperature difference is narrower. Therefore, the metering device must reduce the amount of refrigerant flowing through the system.

In sum, the metering device must always reduce the flow of refrigerant if the outside air temperature increases or if the indoor air temperature decreases. Conversely, it must increase the flow of refrigerant if the outside air temperature decreases or if the indoor air temperature increases. A change in outdoor air temperature varies with weather conditions, and a change in indoor air temperature varies with the internal load (i.e., lights, cooking, people). Of course, proper sizing of the unit compensates for some variations; however, if the unit is undersized or oversized, it cannot be controlled according to the normal settings.

TYPES OF METERING DEVICES

There are several different types of metering devices. They include: hand-operated expansion valve, low-side float, high-side float, automatic expansion valve, thermostatic expansion valve, and capillary tube. The most common devices used in residential air conditioning are the automatic expansion valve, thermostatic expansion valve, and the capillary tube. Each adjusts the refrigerant flow rate, but in quite a different manner.

HAND-OPERATED EXPANSION VALVE

This is basically an orifice in the liquid line. To feed a certain quantity of refrigerant, this valve must be adjusted manually every time the load changes.

LOW-SIDE FLOAT

The low-side float is located in the evaporator, which is in low-pressure side of the refrigeration system. This float helps maintain a fixed level of liquid refrigerant in the evaporator.

HIGH-SIDE FLOAT

As the name implies, this float is located in the high-pressure side of the system. The refrigerant flows from the condenser to the float chamber. As the level in this chamber rises, the high-side float allows the refrigerant to flow into the evaporator.

AUTOMATIC EXPANSION VALVE

This is one of the oldest valves in use, however, it is simple and reliable. This valve is installed at the evaporator inlet. During machine operation, this valve holds pressure at a constant level, regardless of the load. From an energy conservation standpoint, this valve is highly efficient. Expansion valves are discussed in more detail later on in this section.

THERMOSTATIC EXPANSION VALVE

This very common valve meters the flow of refrigerant into the evaporator in direct relation to the rate of evaporation of this liquid in the evaporator, Figure 10-13. This action prevents liquid refrigerant from returning to the compressor. Through this monitoring process, the thermostatic expansion valve maintains constant superheat in the evaporator.

This valve operates in conjunction with several other forces, which help to constantly monitor the temperature of the refrigeration system. One of these forces is a thermal bulb which attaches to the suction line at the evaporator exit. This bulb is filled with refrigerant and senses the temperature at the evaporator exit. Therefore, bulb temperature, evaporator or suction pressure, and superheat spring pressure (a device which ensures the leaving vapor is superheated and contains no liquid) can all affect the thermostatic expansion valve.

Figure 10-13. Thermostatic Expansion Valve. Courtesy, Sporlan Valve Company.

If the load on the evaporator increases, the refrigerant passing through the evaporator picks up more heat. Consequently, the refrigerant at this point is at a higher temperature than the refrigerant at the inlet of the evaporator. The thermostatic expansion valve bulb at the evaporator exit senses this and responds by exerting additional pressure on the valve. The valve then opens, which allows more refrigerant to flow.

If there is a very light load on the evaporator, the liquid admitted to the evaporator does not completely vaporize until it almost reaches the evaporator outlet. Consequently, its temperature and pressure are very close to the temperature and pressure at which it entered. This reduction in temperature is sensed by the feeler bulb, which exerts less pressure on the valve. The valve then tends to close, reducing the flow of liquid. Although the inlet and outlet temperatures and pressures are lower, the difference in pressure remains the same as the superheat setting, so the valve can adjust to this change in load.

In all cases, the thermostatic expansion valve adjusts to changes in load by changes in refrigerant flow. This flow rate then matches the pumping rate of the compressor so that the system is always in balance and maintains its basic capacity.

Checking Superheat. Superheat can be checked by measuring the temperature at the evaporator exit near the bulb and simultaneously checking it at the evaporator entrance. If the evaporator has a manifold, the temperature should be taken at the line beyond the manifold in order to get an accurate reading. Where this type of temperature recorder is not available, the superheat can be determined as follows:

1. Strap a thermometer on the suction line near the bulb from the expansion valve. Be sure to make good contact and pack a bit of Permagum (or other similar substance) around the thermometer bulb to insulate it from ambient air.
2. Attach a gage set to the unit (the method is described in Chapter 11).
3. After the unit runs long enough to stabilize (5 to 10 minutes), read the suction pressure and convert to temperature using the Temperature/Pressure chart , shown in Figure 9-3.
4. Read the temperature at the exit from the evaporator, and calculate the superheat. For example:

Evaporator Exit Temperature:	53 °F
Suction Pressure	70 psig
Saturated Pressure:	41 °F

$$\text{Superheat} = 12\ °F$$

Modern thermostatic expansion valves are quite dependable, and before concluding the valve is bad, other factors in the system which could affect superheat should be checked first.

CAPILLARY TUBE

This is the simplest device used in modern refrigeration, however, its application is limited, as it is not sensitive to system changes. The capillary tube is a tube with a small inside diameter (id), and it divides the high-pressure side of the refrigeration system from the low-pressure side, Figure 10-14. Refrigerant flows, at a predetermined rate, through the capillary tube into the evaporator. This tube is not adjustable and is used only on flooded systems.

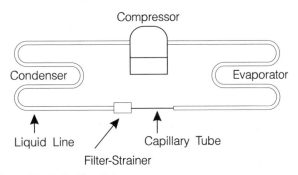

Figure 10-14. Capillary Tube

The capillary tube works because of pressure differences. That is, the small inside diameter of the tube holds back the liquid refrigerant, causing a high pressure situation in the condenser. When the pressure becomes high enough, the capillary tube allows slow flow of the refrigerant into the evaporator. Because of the small id of the capillary tube, it is necessary to keep the system free of dirt and other foreign matter. Otherwise, the capillary tube can become plugged.

The capillary tube is located between the liquid line and the evaporator. The length and diameter of this capillary tube are very carefully matched to a specific unit so the tube can meter the precise amount of refrigerant required by the system. The length of this tube is critical, and it cannot be shortened in the field. The capillary tube often measures 6 to 8 feet. The internal diameter of the capillary tube is in the range of 0.08 to 0.09-inch id.

REVIEW QUESTIONS

1. What is the basic function of a compressor? *MOVE + PRESSURIZE*
2. What is the purpose of an oil separator? *OIL RETURN*
3. Will some oil be carried through the system? Why or why not? *YES. MIXES WITH FREON*
4. What is an important function of the suction gas inside of the compressor shell? *COOLING*
5. Name some possible causes of liquid refrigerant entering the compressor. *OVERCHARGE, RESTRICTED EVAP*
6. What is refrigerant migration? *MOVING TO COMP. OFF CYCLE*
7. What are the three types of condensers? *AIR, WATER, EVAP*
8. What is the function of the condenser? *REMOVE HEAT*
9. What is the function of the evaporator? *ABSORBS HEAT*
10. Name the connecting lines for a refrigeration system.
11. When is a drier installed in the system? *COMPRESSOR BURNOUT*
12. Name the different metering devices. *TXV, AXV, CAP TUB.*
13. What are the two basic areas which affect the ability of the system to increase or decrease capacity? *INDOOR AIR OUTDOOR AIR*
14. If outside air temperature increases, what must the metering device do and why? *REDUCE FLOW. PUMP RATE IS LE*
15. If the outside air temperature decreases, what must the metering device do and why? *INCREASE FLOW, KEEP UP*
16. If the indoor air temperature increases, what must the metering device do and why? *INCREASE*
17. If the indoor air temperature decreases, what must the metering device do and why? *DECREASE*
18. What are the major functions of the thermostatic expansion valve? *METERS BASED ON LOAD*
19. How can superheat be measured? *PRESSURE CHART + THER*
20. What is a capillary tube?

Chapter 11
Cooling Installation

AIR CONDITIONERS

There are several types of central air conditioning systems. They include the single-package unit, the split system, and the fan coil unit. These systems are used for different applications, however it is necessary for a service technician to know how to install all types.

SINGLE-PACKAGE UNIT

A single-package unit is self-contained, meaning that all the necessary components are packaged together, Figure 11-1. A window air conditioner is considered to be a single-package unit; however, there are larger single-package units. This is the least expensive system to install, but the noise of the compressor makes an outside location

preferable. This unit is independent of the heat source, although it can use the same duct system. Often, this type of air conditioner is used when the existing heating equipment does not have the ability to handle the amount of air necessary for cooling. Or, this unit can be used when there is no forced air heating system.

The single-package unit can be used in several different applications. For example, a window air conditioner operates on the air-to-air principle. This means that the air conditioner absorbs heat from the air and rejects the heat back into the air. Another application is air-to-water, in which the air conditioner absorbs heat out of the living space and rejects the heat into water. Other applications include water-to-water and water-to-air. For most residential applications, the air-to-air principle applies.

SPLIT SYSTEM

In the split system, the components are separated: the evaporator is placed inside the house, and the condenser and compressor are located outside. The advantage to this setup is that noise and vibrations are kept out of the conditioned area. A split system is suitable for every kind of home and heating system, Figure 11-2. The condenser is located outside the home in a convenient place and connected by standard-size copper piping to the evaporator cooling coil located at the furnace. The furnace blower is then used for cool air distribution.

Although slightly more expensive than a single-package unit, the split system's specifications automatically eliminate interior noise and the need for large holes to be cut through an exterior wall. Consequently, split systems are the dominant choice for installation in new and existing homes.

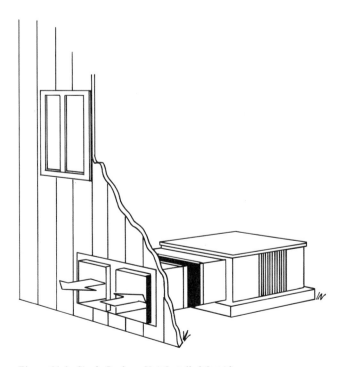

Figure 11-1. Single-Package Unit Installed Outside

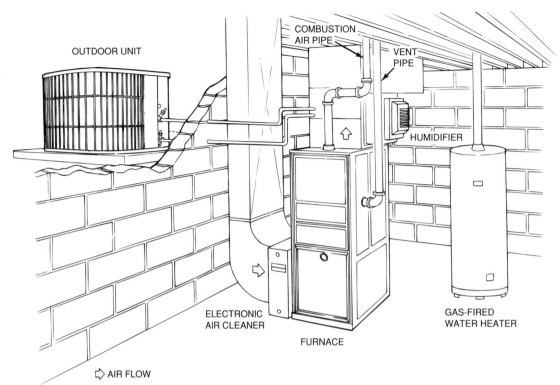

Figure 11-2. *Split System. Courtesy, Carrier Corporation, a Subsidiary of United Technologies Corporation.*

FAN COIL UNIT

An alternate application, which does not use the furnace blower as the prime air mover, uses a fan mounted behind the coil in order to supply the air motion required. The fan coil unit is frequently used in motels, apartments and other places where there is no furnace equipped with a prime air mover and the space for blower and evaporator is limited. This unit does not require a complete duct system, and is used ideally when only one room has to be cooled. Additive heat, in the form of an electric strip heater, can be added to complete the heating-cooling system.

CONDENSER INSTALLATIONS

Several criteria need to be observed when installing the condenser. These criteria include: piping, unrestricted access to air, runoff or ground water, room to service the unit, appearance, and solar influence. Given these criteria, possible locations for the condenser include on a slab on the ground, on a balcony, or on the roof.

PIPING

When selecting a location, it is important to keep in mind the length of the refrigerant lines, as they must be short. The refrigerant lines should also be positioned so that they do not rise any great distance above the evaporator,

because if the lines are positioned vertically, they can trap oil and hinder the operation and efficiency of the system.

UNRESTRICTED ACCESS TO AIR

This is important, because the air discharged from the condenser is hot. If this air bounces off an object and is drawn back into the condenser, poor condenser operating efficiency results.

RUNOFF OR GROUND WATER

The slab on which the condenser is placed should be level, or else runoff water can collect around the unit. Condensers are made to handle rainfall, but direct roof runoff can ruin the fan motor. If the unit is placed in a depression, ground water can flood the unit, creating a short and destroying the wiring.

SERVICE ROOM

It is very important that the service technician have room in which to work around the condenser. If there is improper room around the condenser and the service technician cannot reach a certain part, inadequate service results.

APPEARANCE

The condenser should be placed so as not to detract from the residence. It is also important to place the condenser in a location where noise will not bother the residents.

SOLAR INFLUENCE

It is helpful to place the condenser in a shady location, as direct sun can hinder condenser performance. Direct sun heats up the condenser unnecessarily; however, this is the least important criterion, and the other criteria should take precedence over this one.

CONCRETE SLABS

As mentioned previously, one possible location for a condenser is on a slab on the ground. A slab is necessary to compensate for the vibrations the condenser causes. This slab consists of concrete (approximately 6 inches thick) poured on the ground. This slab should be at least 4 inches above the ground to ensure unrestricted air flow around the base. Some sort of sound-absorbing material should then be placed between the condenser and the slab. A good material to use is a rubber and cork pad. These pads are simple and inexpensive, and they can be cut to the desired size and placed easily between the concrete and the condensing unit.

Most installation instructions recommend maximum distance between the condensing unit and the building. The preferred distance is usually 3 feet or greater from the side of the building and 5 ft minimum from the building roof or overhang. This should allow sufficient space for unrestricted air flow and for access to the service panels, Figure 11-3.

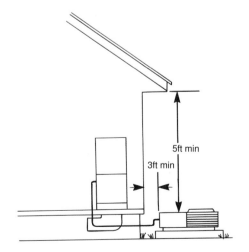

Figure 11-3. Properly-Located Condenser

EVAPORATOR INSTALLATIONS

Evaporator location can sometimes be tricky, due to certain requirements the location must meet. Many times, the evaporator can readily be installed in the furnace. Other times, it may have to be installed in an attic or crawl space. In any event, the evaporator should be located close to the fan, whether the fan is in the furnace or located elsewhere. The supply air plenum can then be installed directly on top of the evaporator, insulated and secured with sheet metal screws. Other points to consider when installing the evaporator are future service and drain piping for the condensate produced by the evaporator.

EVAPORATOR CABINETS

Many furnaces have matching evaporator cabinets, and these cabinets exactly fit the top of the furnace. When this occurs, the evaporator sits directly over the supply air opening. One or two sheet metal screws then secure the evaporator cabinet to the furnace. As this cabinet rests on flanges, it is necessary to insulate the opening between the evaporator cabinet and the flanges with small Fiberglass strips.

ATTIC AND CRAWL SPACES

If an evaporator cabinet is not supplied, other methods must be used. It is very important that the evaporator rest on a solid base, or be suspended securely. In upflow or downflow furnaces, a solid base is often available on which the evaporator can rest. When the evaporator must be mounted vertically, often a wall support is provided, so the evaporator can be placed there. In horizontal situations, the evaporator may be placed in an attic or crawl space.

As with the condensers, it is necessary to control the vibration noises that the evaporator can cause. For this reason, sound-absorbing pads similar to those used with the condensers should be used with the evaporator. These pads should be placed underneath the evaporator, whether it is suspended from ceiling joists or placed on a concrete slab.

SERVICE ACCESSIBILITY

The evaporator requires a service panel in order for the service technician to provide routine maintenance. This rectangular panel is included with the evaporator. The following steps are helpful when installing this access panel:

1. Scribe the outline of the access panel on the supply plenum. Allow about 1/4 to 3/8 inch of metal to extend below the supply plenum, so later it can be screwed into the evaporator cabinet.
2. Cut a hole, approximately 1/2 to 3/4 inch, inside the scribed lines to allow room for attaching the access panel to the supply plenum.
3. Drill holes around the perimeter of the access panel, so it can be tightly secured with sheet metal screws.
4. Drill a 1/4-inch hole (if not already present) directly above the evaporator (in the leaving air stream), and another one directly below (in the entering air stream). These holes are used to measure the temperature drop across the evaporator coil.

DRAIN PIPING FOR CONDENSATE

In normal evaporator operation, approximately 3 pints of condensate per hour are produced in a 1-ton refrigeration system. This amount of condensate increases as the size of the refrigeration system increases. It is necessary for the evaporator to rid itself of this condensate in some manner, so drain piping must be installed. The simplest method to drain the condensate is to attach a plastic hose to the drain connection in the evaporator and direct it toward a floor drain, Figure 11-4. A trap should be installed in the drain line in order to prevent air from being drawn back into the unit. This trap should be at least the size of the drain connection. To ensure proper drainage, it is possible to install the evaporator at a slight pitch.

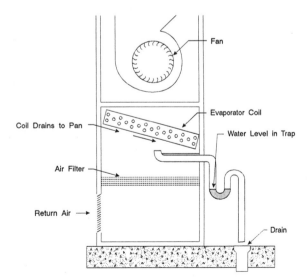

Figure 11-4. Indoor Condensate Drain Piping

If there is no drain close to the evaporator, the condensate must be drained to a different location. For example, the drain line can be run outside of the building, into a dry well. A dry well is nothing more than a hole in the ground

that is filled with rocks or gravel, Figure 11-5. The drain line can be inserted into this dry well (in the middle of the gravel), and the soil then absorbs the condensate. In this situation, however, it is possible for algae to grow in the drain line and eventually plug the drain. For this reason, an auxiliary drain line should also be installed in a location that is readily visible by the home or business owner. This auxiliary drain line can be run to drain to the end of the house (for example, through the end gable of the house). The condensate then falls to the patio or driveway, alerting the owner to call a service technician to fix the primary drain line.

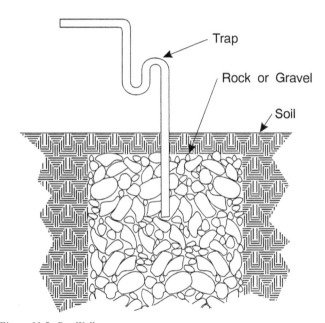

Figure 11-5. Dry Well

REFRIGERANT PIPING

The piping is very important, as it affects the efficiency of the air conditioning system. When connecting the evaporator and the condenser, it is desirable to keep the piping as short as possible. While the following information contains guidelines as to how to install the refrigeration lines, it is very important to follow the manufacturer's recommendations.

Only refrigeration-grade copper piping should be used in air conditioning systems. Air conditioning and refrigeration (ACR) piping is seamless, deoxidized and dried, so it can be used on the job without additional dehydration. While this piping is dehydrated, the system must still be evacuated after installation is complete. This ACR piping, called Type L, is supplied in 50-foot rolls and is available is either soft or hard copper in various sizes. Refrigerant piping is sized by its outside diameter (od).

INSULATION

Suction lines require insulation in order to prevent condensation from forming on the piping and to prevent heat gain, caused by surrounding air. The liquid lines do not necessarily need insulation; however, if they are exposed to high ambient temperatures (i.e., kitchens, boiler rooms), then they, too, should be insulated.

There are several different types of insulation: cork, rock cork, hair felt or wool felt. The types most often used are cork and rock cork, as these are the most durable.

SIZING

Sizing the refrigerant lines can create several problems; however, if the problems are realized from the onset, then sizing becomes easier. When installing refrigerant lines, the following characteristics must be satisfied:

- The correct amount of refrigerant must be fed to the evaporator at all times.
- The refrigerant lines must be large enough to prevent excessive pressure drops.
- Liquid refrigerant must be prevented from reaching the compressor.
- Oil should be returned to the compressor.

The size of the condensing unit and the length of the refrigerant line determine the size of the line. A typical installation requires approximately 30 feet of piping. If the installation requires more than 30 feet, piping of a larger size is required. The length of the run, as well as the size of the line can create a pressure drop through the pipe, which affects system efficiency.

Piping for liquid lines is usually 3/8-inch outside diameter (od) for up to 3 tons, and 1/2-inch od from 3 to 5 tons. The suction line is from 3/4 to 1 1/8-inches od.

Refrigerant piping should never be installed in a cement slab, as this limits access to the refrigerant should a leak be suspected. Also, to ensure good oil return to the compressor, it is important to pitch the horizontal suction lines toward the compressor, approximately 1/8 inch for each 10 feet of piping.

Unsupported refrigerant lines may vibrate and cause noise. Easy-to-use hangers are available, Figure 11-6 to help with this problem. These hangers can be used when the lines run under ceiling or floor joists, and they should be placed approximately every 6 feet.

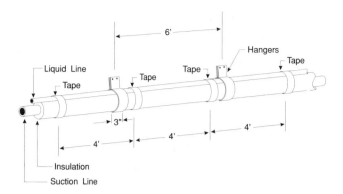

Figure 11-6. *Hangers*

PIPE PACKAGING

Many manufactures prefer to preassemble the refrigerant lines so that the problems of cutting, soldering or flaring are eliminated or reduced in the field. These preassembled packages are called line sets or precharged line sets.

Line Sets. In a line set, the suction line comes already insulated, and the piping is charged with nitrogen. Rubber plugs in the pipe ends hold in this nitrogen. With a line set, the complete system charge is located in the condenser, and the lines must be evacuated and purged after connection. The charge provided is sufficient to operate a certain line length, usually 30 feet.

Specially-designed couplings are found at the ends of the piping in line sets and at the evaporator and condenser. These couplings conserve the charge in the condenser during the connection process. The various fittings are not interchangeable, and the line set must match the fittings provided by the manufacturer of the condensing unit. When the length of the refrigerant line is over 55 ft, the supplier should be notified to determine the proper line size.

After the piping is connected, it is necessary to evacuate, purge, and again evacuate the system. The first evacuation rids the system of all noncondensible gases and moisture. Purging is necessary to rid the piping and evaporator of the nitrogen they contain. The final evacuation is necessary to rid the system of any contaminants found in the purging cycle. This evacuation-purging cycle is described in detail later in this section.

Line sets come in varying lengths (i.e., 10, 20, 30 feet). The length ordered should match closely the distance between the condenser and evaporator. If at all possible, these line sets should not be shortened, as some sets have the capillary built into the liquid line, and shortening the line can destroy operation.

If it is necessary to alter the length of a line set, perform the following steps:

1. Do not remove the rubber plugs. Measure and alter the piping to the proper length. This causes the nitrogen to escape.
2. Perform a leak test on the pipe connections by pressurizing the piping with refrigerant (purging) and checking with a standard leak detector (discussed in more detail later in this section).
3. Evacuate the line set and evaporator using a vacuum pump. As the line set has now been altered, it is necessary to alter the charge to meet manufacturer's guidelines once the system is started.

Precharged Line Sets. Precharged line sets are similar to regular line sets, except they are precharged with the proper amount of refrigerant required for that length. These line sets do not have to be purged or evacuated. Unless the lines are altered, refrigerant does not have to be added. If the lines do need to be altered, it is important to follow the manufacturer's recommendations concerning the procedure and the amount of charge needed for the new line lengths.

To connect a precharged line set, perform the following steps:

1. Route the suction line and liquid line between the evaporator and condenser.
2. Remove the dust caps.
3. Wipe couplings and threaded surfaces with a clean cloth to prevent dirt or foreign material from entering the system.
4. Lubricate the o-rings on each line fitting. These o-rings prevent the refrigerant from leaking out when the piping is connected to the fitting.
5. Attach coupling to threaded surface and tighten fingertight. Once fingertight, use a properly-sized wrench to tighten this connection further. Do not stop tightening until the connection is complete, or else the system charge may be lost. Only tighten until a definite resistance is felt.
6. Proceed to tighten all fittings as needed.
7. Perform a leak test on all connections.

PIPE CONNECTIONS

The refrigerant lines need to be connected in order to form a complete, closed circuit. Using flare connections and brazing are the two methods most frequently used when making pipe connections.

Flare Connections. All fittings have a 45° bevel on the connecting end. Therefore, in order to join the piping and

fittings, a flare connection must be made. This is accomplished by flaring the copper piping at a 45° angle to mate with the fitting, Figure 11-7. The flare nuts draw this flared edge down very tightly to make a completely leak-free joint.

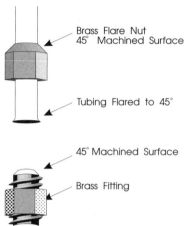

Figure 11-7. Flare Connection

The tools required for making flare connections are a flaring tool which makes the 45° bevel at the end of the copper pipe, Figure 11-8, a pipe cutter, Figure 11-9, and a reamer, Figure 11-10 to remove burrs and ensure a clean surface for the final fitting.

Figure 11-8. Flaring Tool. Courtesy, Imperial Eastman.

Figure 11-9. Pipe Cutter. Courtesy, Imperial Eastman.

Figure 11-10. Reamer. Courtesy, Imperial Eastman.

In order to make a flare connection with piping, perform the following steps:

1. Accurately measure the amount of piping required, allowing for the various fittings that are included in the piping system.
2. Cut the pipe to proper length. In order to keep moisture out of the piping, cut only the amount of piping required for the immediate connection.
3. Use a pipe cutter to ensure the cutoff is square and uniform all around. Position the pipe cutter at the cutoff point and screw it down fingertight. Too much pressure will collapse the side of the pipe, and too little pressure will allow the cutter to wander around the piping, resulting in an elliptical or uneven cut. To produce a clean, square cut, slowly increase the tension as the cutter is rotated.
4. De-burr the pipe to remove the turned-in metal resulting from the cut. Either a reamer or a penknife may be used, but do not use emery cloth or abrasive to remove this type of burr. Also, always hold the open end of the pipe down so that loose burrs do not lodge in the piping and later contaminate the system. Clean all residue from the piping.
5. Clamp the cut end of the pipe in the flaring tool. To ensure proper location, use the automatic gage that is part of the tool.
6. Place a small amount of refrigerant oil on the end of the piping to help make a clean bevel.
7. Rotate the flaring tool to make the 45° bevel.
8. Remove the piping from the block, and assemble the joint.
9. To ensure that this joint is leak-free, use a drop of refrigeration oil on the flare and fingertighten. Then, use two open end wrenches to tighten one-quarter turn only. If a leak occurs, there is sufficient thread to tighten the nut.

Brazing. Another method of joining refrigeration lines is by silver alloy brazing or silver soldering. Silver alloys have a melting temperature of around 1,150 °F, which ensures a strong, reliable and leak-proof joint.

Brazing is used almost exclusively on commercial systems over 5 tons. As such, brazing is seldom used for residential applications. If brazing is required, Modern Soldering and Brazing Techniques is a good reference book to use.

SYSTEM CONNECTIONS

It is important to understand the valves and fittings on the compressor. These valves or fittings are provided on the compressor or line set, making it possible to attach gage manifolds. This allows the service technician to read the operating characteristics of the system. These valves and fittings also make it possible to evacuate and charge the system without losing refrigerant while attaching or disengaging the hoses. A special tool, called a ratchet wrench, Figure 11-11 is used to open and close the service valves. Most compressors are fitted with a service valve on both the suction and discharge sides of the system.

Figure 11-11. Ratchet Wrench with Square Openings. Courtesy, Imperial Eastman.

SERVICE VALVES

Service valves have an inlet, an outlet for the refrigeration lines, and also a smaller gage port. The gage port is used to check operating pressures and temperatures, and to charge and evacuate the system. If the service valve is completely back-seated (stem is turned counterclockwise) the gage port is shut off, Figure 11-12. It is important to note the normal position when system is running.

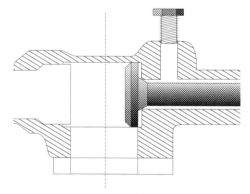

Figure 11-12. Back-Seated Service Valve

After the gage connection is attached to the gage port, the valve stem is turned one to two turns clockwise. This is called cracking the valve, and this allows refrigerant to flow past the valve seat and into the gage port channel, which then registers operating pressures, Figure 11-13. Figure 11-14 shows the valve front-seated by turning it fully clockwise and shutting off the flow of refrigerant through the system.

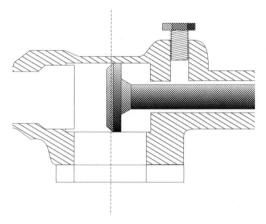

Figure 11-13. *Cracking the Service Valve*

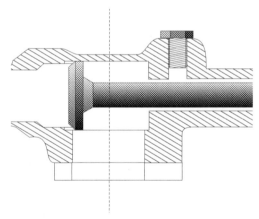

Figure 11-14. *Front-Seated Service Valve*

The gage port has a Schrader valve to hold the charge even when the cap is off, Figure 11-15. Some service valves have two gage ports. The second port is used in factory processing and does not have a valve stem.

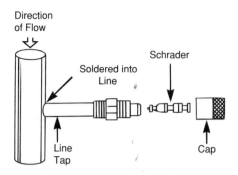

Figure 11-15. *Schrader Valve*

It is a good practice to completely back-seat the service valve before removing the cap to determine whether the port is open or has a stem. A Schrader valve may also be soldered into the line. The Schrader valve functions in exactly the same manner as the service valve gage port; however, it does not have an adjusted valve stem to shut off the system flow. It only has the valve pin and core which must be unseated when the hose is attached.

If the manifold hoses do not have an unseated pin, then a second device must be used: an unseating coupler. This attaches to the end of the hose connection and includes the pin needed to depress the valve core on the Schrader fitting. Some units have both service valves and one or more Schrader fitting. There is a tool, called a valve core remover/installer, Figure 11-16, which unseats the valve and pulls the stem out of the way. This allows more flow through the gage port and can save considerable time in evacuating and charging the system.

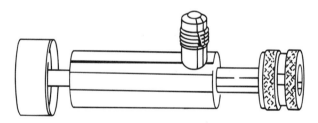

Figure 11-16. *Valve Core Remover/Installer*

CHARGING MANIFOLD

The standard tool for servicing/charging air conditioning equipment is a charging manifold, Figure 11-17. This manifold has been developed to perform a number of different jobs without changing hose connections into the system. The charging manifold has a compound or low pressure gage which has scale from 250 psi down to 30 inches of mercury vacuum; therefore, it can read vacuum and positive pressures. A second gage is normally located to the right of the compound gage. This is a high-pressure gage and should read at least to 400 psig.

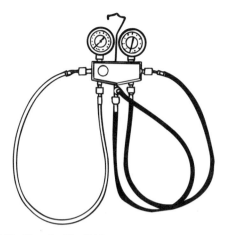

Figure 11-17. *Charging Manifold*

When the gage manifold valves are front-seated, the center port is completely closed off from the gages and gage ports, Figure 11-18. The gage port, however, is open to the system and the operating pressures are indicated. Opening either valve opens that side of the system to the center port. Three hoses, each of a different color, are attached to this manifold: one on the high side (red), one on the low side (blue), and the third hose in the center (yellow).

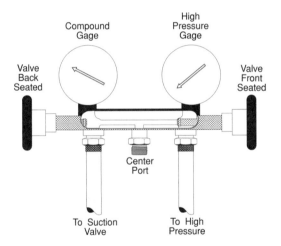

Figure 11-18. *Center Port Closed From Gages*

The hose to the compound gage attaches to the suction service valve gage port. The hose to the high-pressure gage attaches to the discharge or liquid service valve gage port or liquid line Schrader fitting. The center hose is attached to either the refrigerant supply cylinder or the vacuum pump.

There are charging stations, Figure 11-19 available which include additional piping and hand valves. When using a charging station, vacuum pump, refrigerant cylinder or charging cylinder can be selected merely by turning a valve, without changing the hose from one to the other. The hand valves on the gage manifold open each side of

the system to the center hose and, if both are open, both sides are open to the center base. These valves should be closed to ensure no refrigerant leaks out of the system.

Figure 11-19. *Charging Station. Courtesy, Robinair Division, SPX Corporation.*

ATTACHING THE MANIFOLD TO THE SYSTEM

To attach the manifold to the system, the center hose from the gage manifold should be connected to the vacuum pump, Figure 11-20. The fitting is screwed down fingertight. The compound gage hose connects to the gage port on the suction valve of the compressor and is screwed down tightly. The high-pressure gage hose connects to the gage port of the liquid line valve or Schrader valve in the liquid line and is screwed down tightly. The compound gage now shows vacuum or pressure on the suction side of the system, and the high-pressure gage shows pressure on the discharge side, Figure 11-21.

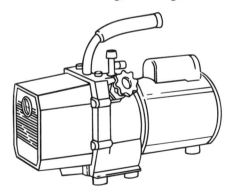

Figure 11-20. *Vacuum Pump*

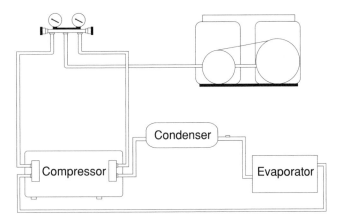

Figure 11-21. *Connecting the Charging Manifold Prior to Evacuation*

This procedure for attaching hoses is satisfactory on initial installation, prior to the time the system has been evacuated. However, it is not satisfactory after system evacuation, due to the possibility of getting air or moisture into the system. The proper procedure for attaching the hoses to the evacuated system is as follows:

1. Unscrew and remove the hoses to the suction and discharge valves prior to attaching the refrigerant drum.
2. Attach the center hose to the refrigerant drum and close both manifold valves.
3. Crack the refrigerant drum valve to establish pressure in the hose.
4. Attach the hose from the discharge side of the system very loosely to the discharge gage connection.
5. Crack the hand valve on the discharge side of the manifold. This allows some refrigerant to flow through the hose and purge any air in the line.
6. After the line is purged, screw down the hose quickly and securely against the gage port. Then, repeat this procedure on the suction side, so both hoses are completely purged of air and moisture before being opened into the refrigeration system.
7. When attaching the hoses, make sure the service valve is completely back-seated, so the gage port does not open into the system.
8. Once a system is charged and under pressure, quickly remove the hoses, or else the charge can be lost. To remove the hose, hold the metal angle tightly against the gage port as the nut is loosened. When the nut is free, remove quickly so no charge is lost. The valve should be back-seated when removing the hose to prevent loss of charge.

EVACUATING, PURGING AND LEAK TESTING

All new installations must be evacuated to a deep vacuum in order that all noncondensible gases and moisture are removed prior to charging the system. As explained previously, the presence of moisture in a system can render it inoperable in a very short time. Proper evacuation assures a dry, uncontaminated system. Purging then rids the system of the nitrogen found in the piping and evaporator. During the evacuating and purging, it is also necessary to check for leaks, to ensure all piping connections are sound.

EVACUATING AND PURGING

Vacuum is measured in inches of mercury. The absolute possible vacuum is 29.999 inches, referred to as 30 inches barometer, 0.001 inches mercury absolute, or 25 microns. To remove all noncondensible gases, the vacuum must be 29-1/4 inches of mercury on the system. During evacuation, a very accurate U-tube manometer or electronic micron gage should measure the vacuum. However, most vacuum pumps only have one inlet connection and no provision for attaching the manometer. To solve this problem, in a two-stage vacuum pump, a small T connection should be inserted at the inlet to the pump with one leg connecting the manometer, and the other legs connecting with the system. If a three-stage pump is used, a manometer is not required, as this type of pump pulls a deeper vacuum.

To evacuate and purge a system, perform the following steps:

1. Start the vacuum pump once the pump is connected and both manifold service valves are open.
2. Stop the vacuum pump at least once during operation and check that the vacuum is holding. If it is not, there is a leak in the system which must be located and repaired before an absolute vacuum can be obtained.
3. Turn off the pump and front-seat the manifold valves.
4. Remove the hose from the vacuum pump and attach it to the refrigerant drum.
5. Open the refrigerant drum and manifold valves and allow the system to pressurize to 2 or 3 lbs (this purges the system). Check for leaks, and repair any that are found.
6. Close both manifold valves, disconnect the refrigerant drum from the center port and open the suction line valve slowly, purging the system down to 0 psig.
7. Evacuate the system a second time through the suction service valve, until the vacuum within the system does not rise above 29.7 inches of mercury.

TESTING FOR LEAKS

As stated previously, all new installations should be checked for leaks during the evacuation process. It is easiest to check during evacuation, because a leak in the system allows air to enter and raise the pressure when the system is under a vacuum. If a leak is detected and the system loses most of its charge, 150 psig of refrigerant should be introduced into the system. Then, the system should be checked for leaks with a halide torch or an electronic leak detector. Electronic leak detectors work well, because they can sense extremely small refrigerant leaks.

If a system has been operating for some time, the first leak inspection is usually a visual one. For example, traces of oil at any joint or connection should lead the service technician to suspect that refrigerant is leaking at that point, because refrigerant contains a small quantity of oil.

The most efficient leak detector is the electronic detector, Figure 11-22. These detectors contain an element that is sensitive to halogen gases (i.e., R-12, R-22). When the detector is turned on and the probe passed around all the fittings and connections of the system, a slow ticking noise can be heard. If a leak is detected, the ticking becomes loud and rapid.

Figure 11-22. Electronic Leak Detector. Courtesy, Carrier Corporation, a Subsidiary of United Technologies Corporation.

Another method of checking leaks is to use a halide torch, Figure 11-23. The flame changes color in the presence of a halogen gas, so if refrigerant is present, the flame turns from blue to green. When using a halide torch, it is advisable to turn the flame down as low as possible and make sure the area is well ventilated.

Figure 11-23. Halide Torch

One rather simple and inexpensive method to check for leaks is to use a soap solution and apply it to all joints and connections. A typical leak causes soap bubbles to form around its opening.

If no leaks are found and the system contains the correct charge, the system can be readied for normal operation. This is accomplished by back-seating the compressor service valve and closing the gage manifold hand valves. Then, the hoses should be removed quickly from the compressor (so as not to lose charge), and the gage caps replaced.

REPAIRING THE LEAK

If a leak is found, it must be repaired before proceeding with final charging and operation. In order to repair a leak, the system must be pumped down.

Many systems have valving arrangements which allow repair of leaks in the low-side piping and evaporator; or removal of the expansion valve or drier; or complete replacement of the compressor, without losing the refrigerant charge. This is accomplished by pumping the entire charge into the condenser and holding it there. This is called pumping down the system.

In order to pump down the system without losing the charge, the system must be equipped with a suction service valve, a discharge service valve, and a liquid valve. Some line sets cannot be changed without losing the charge, and precharged line sets must have a multiple shot coupling if the charge is to be saved. The manufacturer's literature should be checked for details concerning saving the charge when repairing leaks.

The system must also have an adjustable low pressure control which can be turned down to zero prior to pumpdown. Nonadjustable low-pressure controls should be set at about 5 lbs so the compressor can be pumped down to this point before stopping. This allows some charge to be saved, but the balance must be bled out. The low pressure control can be jumpered to start the compressor, however, jumpering is not recommended to keep the compressor running.

To pump down the system, perform the following steps:

1. Turn off the unit disconnect.
2. Set the room thermostat well above room temperature.
3. Place the fan control in the on or cont. position.
4. Set the low pressure control at zero.
5. Attach a gage manifold to the unit.
6. Turn the suction and discharge valves clockwise, one to two turns, in order to open the gage ports.
7. Completely front-seat (turn clockwise) the liquid line valve. (In a precharged line set, disconnect the multiple shot.)
8. Start the compressor by the unit disconnect and allow it to run until suction pressure comes close to zero, then stop the compressor.
9. After the first pump down cycle, the pressure usually increases. Start the compressor again (jumper the low pressure switch, if necessary, to start and repeat the pump down cycle until the suction pressure holds at 1 or 2 psig). A slight positive pressure in the system prevents air and moisture from being drawn in if a leak exists in the evaporator.
10. Completely front-seat the discharge service valve.
11. Repair any leaks in the compressor or low side, or change necessary components.
12. Evacuate the system and install a new drier if the system has been opened.
13. Open suction and discharge valves (turn counterclockwise) and slowly open the liquid line valve. (In a precharged line set, reconnect the quick connects.)
14. Reset the low temperature control.
15. Check operating temperatures and pressure. Add charge if needed.

CHARGING THE SYSTEM

Charging the system simply refers to adding refrigerant to an air conditioning system once all the components and piping are connected. This is not as easy as it sounds, for every component in the system must receive the correct amount of refrigerant, or else the system may fail to operate.

Refrigerant may be added to the system in either vapor form or liquid form, depending upon the amount of charge

required, the valving arrangement on the unit and the outside temperature. For new installations, or where a large amount of charge is required, the system is normally charged with faster-acting liquid. If only a small quantity is required, then the system is normally charged with vapor through the suction side. Some units, because of their valving arrangement, can be charged only in liquid or only in vapor form and this will determine how the system is charged.

It is also important to keep in mind that the refrigerant drum contains both vapor and liquid, with vapor at the top of the drum and liquid at the bottom. As outside temperature increases, the amount of vapor in the drum also increases. If the temperature drops, the amount of liquid in the drum increases. Therefore, it may be necessary to either heat or cool the drum with water in order to obtain the correct amount of gas or liquid required for charging. Never use an open flame to warm the refrigerant drum, as rapid expansion of vapor can rupture the tank. If more vapor is required, the refrigerant drum should be placed upright. If more liquid is required, the refrigerant drum should be turned upside down.

CHARGING WITH VAPOR REFRIGERANT

To accomplish vapor charging, it is necessary to allow the vapor to move out of the refrigerant drum into the low-pressure side of the system. When the system is not operating, vapor may be added to either the low- or high-pressure sides of the system. However, when the system is operating, vapor may only be added to the low-pressure side.

When the outside temperature is warm, it is easy to use vapor refrigerant to charge a system, as the pressure in the low-pressure side of the system is lower than the pressure in the drum, allowing the vapor to move freely. When the outside temperature turns colder, however, it is necessary to warm the refrigerant drum, otherwise, the pressure in the drum is lower than that in the system, and the vapor cannot move freely. As stated previously, it is important to never use an open flame to warm the refrigerant drum. Immersing the drum in warm water is much safer.

To charge the system with vapor refrigerant, perform the following steps:

1. Purge the compound gage hose and attach it to the suction service valve.
2. Purge the high-pressure gage and attach it to the discharge service valve.
3. Connect the refrigerant drum to the center hose and charge vapor refrigerant through the low-pressure or suction side of the system, Figure 11-24.

4. To accelerate charge, front-seat the suction service valve until there is approximately a full charge (the full charge can be determined by checking the manufacturer specifications). At that time, open the service valve, start the compressor, and run it for about 5 minutes, to allow the system to stabilize.

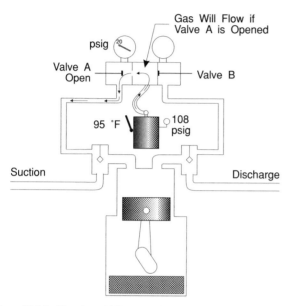

Figure 11-24. *Charging with Vapor Refrigerant*

CHARGING WITH LIQUID REFRIGERANT

The liquid line is normally used when charging a system with liquid refrigerant. It is very important that liquid refrigerant not enter the compressor, therefore, charging liquid refrigerant into the suction line of the compressor is not recommended.

When using liquid to charge a system, the manifold may be connected in the same way as for vapor, but the refrigerant drum is turned upside down and the high-pressure side of the manifold is opened to the refrigerant drum, Figure 11-25. Again, when a complete charge is needed, the liquid line valve can be front-seated until the charging is almost complete.

WEIGHING THE REFRIGERANT

Sometimes it is necessary to add only a certain amount of liquid refrigerant to a system. This is accomplished by using a very accurate scale, as the charge in many systems cannot vary more than 1 or 2 ounces.

Dial scales are sometimes used, Figure 11-26, however calculating the final drum weight, after the prescribed amount of refrigerant moves from the drum into the system, can be confusing. For this reason, electronic scales are often used, Figure 11-27. These scales work very well, because they can be adjusted to zero when a full drum is placed on them. This scale keeps track of the amount of refrigerant flowing into the system and records this amount on a digital screen. The service technician can then turn off the refrigerant when the amount needed shows up on the digital screen.

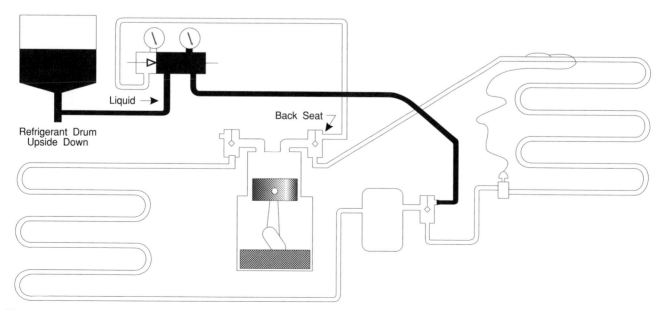

Figure 11-25. *Charging with Liquid Refrigerant*

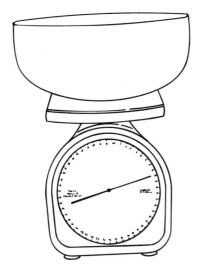

Figure 11-26. Dial Scale

Figure 11-27. Electronic Scales Used to Measure Refrigerant. Courtesy Robinair Division, SPX Corporation.

WARNING: To prevent personal injury, wear safety glasses and gloves when handling refrigerant. Compressor flooding can result if the system is overcharged.

Graduated cylinders, called charging cylinders, can also be used to measure the refrigerant, Figure 11-28. These cylinders are accurate to ± 1/4 oz. Instructions come with the unit, and the service technician need only to fill the cylinder with the refrigerant. The amount added to the cylinder is visible, and the pressure gage on top of the cylinder shows the temperature of the refrigerant. It is important to know the temperature, as liquid refrigerant can be of different volumes depending on the temperature. The cylinder then controls the refrigerant as it is fed into the system, via the gage manifold.

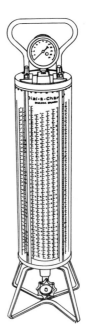

Figure 11-28. Charging Cylinder

When using a weighed amount of refrigerant, the attachments to the suction and discharge service valves are made in the same manner as with vapor charging, and the suction service valve is front-seated. Initially, the charge is allowed to enter the system from the pressure of the refrigerant in the drum. As the pressure decreases, the charging process can be increased by starting the compressor.

CHECKING REFRIGERANT CHARGE

The most accurate and widely used method of checking refrigerant charge is to use a chart which relates suction pressure to outside temperature and discharge pressure, Figure 11-29. The outside temperature is normally indicated by a curve down through the center of the chart. When the system is charged correctly, the suction and discharge pressures a service technician measures should be very close to the points noted on the chart.

Another method of checking charge is to use a sight glass located in the liquid line between the condenser and the drier, Figure 11-30. A sight glass is a small glass pane through which refrigerant can be observed as it passes through the liquid line. After about 5 minutes operating time, the refrigerant should be almost 100 percent liquid at the point where it passes through the sight glass. If bubbles are observed in the line, the system is undercharged and refrigerant must be added (preferably in vapor form). The refrigerant in the sight glass should be free of bubbles when the system is properly charged.

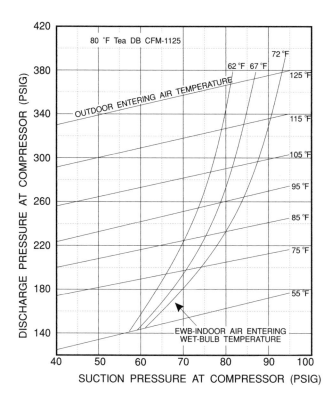

Figure 11-29. *System Charging Chart*

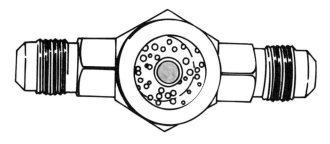

Figure 11-30. *Sight Glass*

Overcharge of a system can be indicated by high head (discharge pressure) or high suction pressure, or it may be indicated by sweat on the suction line. If the system becomes overcharged, the following steps should be performed:

1. Turn off the valve on the refrigerant tank.
2. Remove the hose from the tank and slightly crack the low-side manifold valve. This allows the refrigerant to bleed out of the center hose. Bleed a little bit of refrigerant at a time.
3. Close the manifold valve for system to stabilize. After a few minutes, check that the pressures and temperatures are corrected. If necessary, continue until correct charge is obtained.

FIELD WIRING

When installing an air conditioning system, it is important for the service technician to understand a few guidelines to ensure the power supply is safe. Often, the line-voltage power supply is installed by an electrician, however, the service technician often installs the control voltage for the thermostat. The manufacturer provides step-by-step installation instructions for field wiring connections. It is important to follow these instructions.

The field wiring of the condenser and thermostat for original installation consists of:

- a disconnect switch box at the unit so power may be terminated if needed.
- line voltage wire from the main circuit breaker or electrical entrance to the disconnect.
- line voltage wire from the disconnect to the compressor contactor.
- two wires from the thermostat to the blower relay and contactor.
- line voltage wires from the source to the blower relay and from blower relay to fan control and/or blower motor.
- low voltage wires from thermostat to blower relay and from blower relay to the compressor contactor.

The installation instructions and wiring diagram describe the sizing and fusing of the disconnect switch. The disconnect switch is normally not supplied with the unit and must be purchased separately. The safety disconnect box must be installed in accordance with the National Electrical Code (NEC) and other any local codes or ordinances that apply. The disconnect switch box can be located on the condenser. For 240-vac, single-phase supply, a three-lead wire (two hot and one ground) is run from the main service entrance to the line side of the disconnect switch.

NOTE: The main circuit breaker should be off, as well as the disconnect switch, and no power should be applied to the unit until all wiring is in place. The manufacturer's instructions should advise the application of power.

When adding cooling to an existing heating-only system, it is necessary to replace the thermostat with a fan control. This thermostat has a Y terminal for cooling and a G terminal for the fan control. It is also necessary to provide a blower relay (the furnace may already provide this). This blower relay is a single-pole, double-throw (SPDT) relay with one set of normally-open contacts (NO) and one set of normally- closed (NC) contacts, plus a coil. This relay allows the blower to cycle with a compressor on cooling and in the case of multispeed motors, allows the selection of a higher motor speed for cooling than for heating.

SYSTEM ADJUSTMENTS

In order for any air conditioning system to operate properly, it must have the correct amount of air passing over the evaporator coil. The methods of changing the amount of air delivered by the blower have already been discussed. These included adjusting the pulley on a belt-drive blower or changing either the speed controller or tap-wound motor connections on a direct drive blower.

The amount of air passing over the coil can be measured by its static pressure drop. Measuring the temperature drop across the coil determines whether or not this air is sufficient. All air conditioning systems are assumed to deliver between 400 and 450 cfm of air per ton of rated refrigeration capacity. In other words, a 2-ton system should have between 800 and 900 cfm of air across the coil, and a 3-ton system between 1200 and 1350 cfm of air across the coil.

The amount of air across the coil and its temperature drop are related but do not correlate exactly. This is because the temperature drop across the coil can be affected by humidity. For instance, if the basement has a high humidity level, it is difficult to get the proper temperature drop across the coil, even with the correct amount of air. This is because the coil is using much of its capacity to reduce the moisture content rather than reducing the sensible temperature of the air. Under these circumstances, the coil is actually functioning correctly even if the temperature drop is less than expected. Removal of humidity is just as important to the comfort conditions in the house as reducing the sensible temperature of the delivered air. Thus, the temperature drop across the coil can vary quite a few degrees. Under normal conditions, the temperature drop across the coil is between 18 and 20 °F.

STATIC PRESSURE DROP

When discussing evaporator installation, it was mentioned that most units have 1/4-inch holes drilled in the plenum, one directly above the evaporator coil and one directly below the evaporator coil. These are for measuring the entering and exiting air temperatures or for measuring the static pressure across the evaporator coil. These holes may be covered up by insulation on the inside. If so, an awl or sharp-pointed device can be inserted into the hole to clear it for insertion of the instruments. If these holes are not present, they can be drilled. However, the service technician should be extremely careful not to drill into the evaporator. The hole should be located on the center line of the evaporator.

The pressure drop across the coil, or the static pressure, is measured in inches W.C. with an inclined manometer or draft gage. Static pressure measured across a dry coil, that is, when the refrigeration circuit is not operating, is most accurate. However, many manufacturers give the range of static pressure over a wet coil, and some give both. If the refrigerant system has been in operation, it is better to wait 10 to 15 minutes with the blower on before taking these measurements for a dry coil.

An inclined manometer, Figure 11-31 allows direct reading of static pressure across a coil. To take a static pressure reading with an inclined manometer, the following steps should be performed:

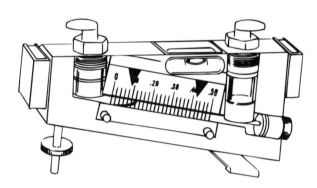

Figure 11-31. Inclined Manometer

1. Level the manometer with either the adjustable foot or by placing it against a steel surface and holding it with its magnet. A spirit level is built into the manometer for leveling.
2. Open each connector for the piping by turning one turn counterclockwise.
3. Adjust the scale so that the zero reading is exactly at the point where the fluid in the manometer shows.
4. Attach one hose to the connector at the lower part of the manometer, and place the other end in the outlet end of the evaporator. Attach the second hose to the connector at the higher end of the manometer, and attach the opposite end to the lower hole in the evaporator. Tape over tube and holes on the evaporator.
5. Turn on the blower (reduce thermostat setting well below room temperature to keep air conditioning system off if the measurement is to be taken on a dry coil) and measure the static pressure with the manometer.

Many manometers have a mirror behind the fluid to allow accurate reading in the scale by aligning the mirror image and the actual fluid, thus, eliminating parallax. The reading in inches W.C. can then be compared to the manufacturer's specifications which show the amount of cfm passing across the coil, Figure 11-32.

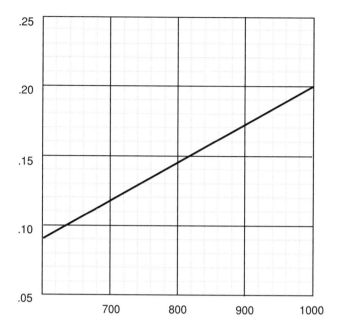

Figure 11-32. Static Pressure Chart

The manufacturer's specifications show readings for specific coils and the proper coil must be selected. Note whether the values given are for a wet or dry coil. The readings are often given in a spread (0.20 to 0.22), as it is usually difficult to read an exact number with these small increments.

If the reading obtained on the manometer is below the readings given in the table for the desired cfm, then the blower speed should be increased and the pressure tested again. If the reading is above the range selected, then the blower speed is greater than desired and should be reduced.

The amperage draw of the motor should be also be checked, when it is operating at a speed that produces the required static pressure. This reading should then be compared with the rated amperage on the motor nameplate. If the reading is close to the amperage on the nameplate, then it can be assumed that the motor is not overloaded at the speed selected for proper air over the evaporator.

TEMPERATURE DROP

In order to check the temperature difference or temperature drop across the coil, perform the following steps:

1. Turn on the unit and allow it to operate for 5 to 10 minutes or until the system stabilizes.
2. If two thermometers are available, place one in each of the two holes used for the static pressure check. If only one thermometer is available, take the reading

on the entering side of the coil and then move the thermometer to the other hole and take a reading on the leaving side. If more convenient, measure the entering air in the return air duct, close to the furnace.

A properly operating coil, with the required amount of air across it, has a temperature difference between these two readings of about 18 °F. As mentioned earlier, the temperature drop can vary with the humidity in the immediate area. Therefore, the system is not necessarily malfunctioning if the temperature drop is less than 18 °F.

In general, if the temperature difference is lower than 18 °F, then the blower speed is too fast. If it is higher than 18 °F, the blower speed is too slow. However, this should be considered in view of the results of the static pressure test. If the blower speed is about right, the lower temperature drop is likely due to high humidity. There may, however, be something else wrong in the refrigeration system, and checks for this are discussed later.

MAINTENANCE

Basic maintenance for blower motors, including adjustment and lubrication, has been covered in previous chapters; however, the same procedures apply to a cooling system check. Basically, the key to an efficient air conditioner is to keep it clean.

MAINTENANCE CHECK LIST

Perform the following check list to ensure proper operation of the cooling system:

- Check the evaporator coil for any clogging due to dirt or lint. If there is accumulation, spray the coil with coil cleaner and then clean with a wire brush.
- Check the condensate drain for free flow by pouring a cup of water down the drain. Make sure the water runs freely.
- Check all piping joints and connections for signs of oil, etc.
- Check and change the furnace filter once a month, regardless.
- Check that the condenser is clean, especially prior to start up. Suggest a cover be used during winter months. Oil the condenser once a year with a few drops of SAE No. 10 nondetergent oil.
- Check the main breaker switches and check all electrical connections for tightness. Observe wire insulation for fray or possible over heating.

REVIEW QUESTIONS

1. What components make up the low-pressure side of a refrigeration system? EVAPORATOR

2. What components make up the high-pressure side of a refrigeration system? CONDENSER

3. What is a self-contained system? PACKAGED

4. Where can condensers be located? OUTSIDE SLABS

5. How are condensers mounted on the ground?

6. Why is a condensate trap important?

7. How should horizontal suction lines be pitched and why? TOWARDS COMPRESSOR FOR OIL RETURN

8. When is an oil trap necessary in the system? EVAP BELOW COMP.

9. What kind of piping is used in air conditioning? ACR

10. Name the tools required for making a flare connection. TUBE CUT, REAMER, FLARING TOOL

11. Where does a precharged system store the refrigerant charge? IN THE CONDENSER

12. What electrical connections are required for the installation of a condenser? LINE + CONTROL

13. When adding cooling to an existing heating system, what additional parts have to be supplied? FAN CONTROL TSTAT FAN RELAY

14. What are the basic service valves on a compressor and where are they located? NEAR COMPRESSOR

15. Back-seating a service valve does what? CLOSES OFF GAGE

16. What is meant by cracking the valve? OPEN ED

17. When the valve is fully front-seated, what is the situation as far as the system is concerned? COMPRESSOR ISOLATED

18. What is a Schrader valve? SERVICE VALVE WITH NO HANDLE

19. What is a charging manifold?

20. When the gage manifold valves are turned full clockwise, is the center port open or closed? Are the gage ports open or closed? CLOSED, OPEN

21. Where is the refrigerant drum attached? CENTER PORT

22. Where is the vacuum pump attached? CENTER PORT

23. What must be done to the hoses prior to attaching them to the system? Why is this done? PURGE, KEEP SYSTEM CLEAN

24. Why is the system evacuated? CLEAN MOISTURE + CONTAMINANTS

25. Why is it necessary to draw a deep vacuum? BOIL THE WATER

26. How is the vacuum measured during evacuation? MICRON GAGE

27. What would be the normal way of charging a new system?

28. If only a small amount of charge is required for an existing system, would vapor or liquid be used? Why? VAPOR, MORE CONTROL ACCURACY

29. How can the proportion of gas to liquid in the refrigerant cylinder be changed?

30. Name three ways of checking for the proper refrigerant charge in a system. PRESSURE CHART, SITE GLASS SUPERHEAT

31. Name four ways to check the system for leaks. HALIDE, BUBBLE, ELECTRONIC, DYES

32. What is meant by system pump down? EVACUATING

33. How much air per ton is required for proper operation? 400 CU FT/MIN

34. What is a normal temperature drop across the coil? 20°

35. How is static pressure measured? MANOMETER

36. What information can be obtained from the static pressure reading?

37. List the major maintenance items for a refrigeration system. FILTERS, COIL RESTRICTIONS, CONDENSATE LINES, ELECTRICAL CONNECTIONS, FAN LUBE OIL RESIDUE AT JOINT

Chapter 12

Humidification

Humidification is a very important aspect in heating and air conditioning. When a given space has too much or too little humidity (moisture in the air), the people or objects within the space can be adversely affected. For example, low moisture content destroys the internal protective barriers of the throat and nasal passages, causing them to dry out. When these internal barriers are destroyed, bacteria and viruses can lodge in the throat and nasal passages, resulting in increased illness.

Low humidity can also ruin household furnishings. Most materials, especially wood, paper and rugs, absorb or lose water depending upon the moisture content of the surrounding air. They shrink as they dry out, and swell as they absorb water. Excessively dry air can cause furniture and woodwork to crack or warp and rugs to wear out much more quickly, because the fibers dry and break. Another common symptom of low humidity is static electricity, which gives one a shock when an object is touched.

On the other hand, high moisture content causes furniture to swell, doors to stick or not close properly, and water to condense on windows. This condensate then runs, ruining windowsills, drapes, walls and rugs. Clearly, it is important for a total comfort system to control the moisture in the air within acceptable limits during both the heating and cooling seasons.

MEASUREMENT

Moisture in the air is called humidity. The ability of the air to hold moisture varies with its temperature. When air is warmer, it can hold more moisture; when air is cooler, it holds less moisture. Relative humidity is based on a comparison between the amount of moisture actually in the air and the maximum amount of moisture that air can hold under the same conditions.

Humidity is measured in grains per pound of dry air. It takes 7000 grains to equal one pound. When 100 percent saturated, air at 70 °F can hold 110 grains of moisture per pound of air. If the same air at 70 °F only contains 55 grains of moisture per pound, then its relative humidity is 50 percent (55 ÷ 110 = 0.50, or 50 percent). Most people feel comfortable at 30 to 60 percent relative humidity.

As air temperature increases, its ability to hold moisture also increases. Consequently, if the same 70 °F air is heated to 85 °F, it can now hold 185 grains of moisture. If no moisture is added to the air, then its relative humidity drops to 30 percent (55 ÷ 185 = 0.297, or 30 percent). Thus, increasing the temperature of the air, without adding more moisture, reduces the relative humidity.

Conversely, if air temperature is reduced, its ability to hold moisture decreases, and the relative humidity increases. For example, cooling the 70 °F air down to 55 °F without adding moisture increases the relative humidity to 85 percent. This is because 55 °F air, when saturated, can only hold 65 grains of moisture (55 ÷ 65 = 0.846, or 85 percent). Since moisture content is dependent upon the temperature of the air, it is necessary to state it as relative humidity.

WET-BULB AND DRY-BULB TEMPERATURES

Air moisture content can be measured by using a combination of wet- and dry-bulb temperatures. The wet-bulb temperature measures the humidity in the air and is taken with a thermometer whose bulb is covered with a cloth soaked in distilled water. The temperature is taken after the thermometer is exposed to a strong air stream. Dry-bulb temperature is the actual temperature of the air, or sensible heat level, of the air. Dry-bulb temperature is higher than the wet-bulb temperature until 100 percent relative humidity (air saturation) is reached, at which point they are equal.

db temp.	WB DEPRESSION																													
	1	2	3	4	5	6	7	8	9	10	11	12	13	14	15	16	17	18	19	20	21	22	23	24	25	26	27	28	29	30
32	90	79	69	60	50	41	31	22	13	4																				
36	91	82	73	65	56	48	39	31	23	14	6																			
40	92	84	76	68	61	53	46	38	31	23	16	9	2																	
44	93	85	78	71	64	57	51	44	37	31	24	18	12	5																
48	93	87	80	73	67	60	54	48	42	36	34	25	19	14	8															
52	94	88	81	75	69	63	58	52	46	41	36	30	25	20	15	10	6	0												
56	94	88	82	77	71	66	61	55	50	45	40	35	34	26	24	17	12	8	4											
60	94	89	84	78	73	68	63	58	53	49	44	40	35	31	27	22	18	14	6	2										
64	95	90	85	79	75	70	66	61	56	52	48	43	39	35	34	27	23	20	16	12	9									
68	95	90	85	81	76	72	67	63	59	55	51	47	43	39	35	31	28	24	21	17	14									
72	95	91	86	82	78	73	69	65	61	57	53	49	46	42	39	35	32	28	25	22	19									
76	96	91	87	83	78	74	70	67	63	59	55	52	48	45	42	38	35	32	29	26	23									
80	96	91	87	83	79	76	72	68	64	61	57	54	54	47	44	41	38	35	32	29	27	24	21	18	16	13	11	8	6	1
84	96	92	88	84	80	77	73	70	66	63	59	56	53	50	47	44	41	38	35	32	30	27	25	22	20	17	15	12	10	8
88	96	92	88	85	81	78	74	71	57	64	61	58	55	52	49	46	43	41	38	35	33	30	28	25	23	21	18	16	14	12
92	96	92	89	85	82	78	75	72	69	65	62	59	57	54	51	48	45	43	40	38	35	33	30	28	26	24	22	19	17	15
96	96	93	89	86	82	79	76	73	70	67	74	61	58	55	53	50	47	45	42	40	37	35	33	31	29	26	24	22	20	18
100	96	93	90	86	83	80	77	74	71	68	65	62	59	57	54	52	49	47	44	42	40	37	35	33	31	29	27	25	23	21
104	97	93	90	87	84	80	77	74	72	69	66	63	61	58	56	53	51	48	46	44	41	39	37	35	33	31	29	27	25	24
108	97	93	90	87	84	81	78	75	72	70	67	64	62	59	57	54	52	50	47	45	43	41	39	37	35	33	31	29	28	26

Figure 12-1. Wet-Bulb Depression Chart

The difference between the wet- and dry-bulb temperatures is called the wet-bulb depression. Calculating this depression allows the relative humidity to be calculated. Figure 12-1 shows a chart of wet-bulb depressions. Using the chart, when the temperature in a room measures 60 °F (dry-bulb temperature), and the wet-bulb temperature is 50 °F, the depression is 10 °F. The chart shows that the relative humidity is 49 percent. If the wet-bulb depression is 0 °F, the relative humidity is 100 percent, meaning the air is saturated with moisture.

A sling psychrometer is a simple instrument that simultaneously measures the wet- and dry-bulb temperatures, Figure 12-2. It consists of two thermometers, mounted side-by-side, on a rotating arm. A cloth, or wick, is soaked with distilled water and placed around the bulb of one of the thermometers. The psychrometer is held by the handle and the two thermometers are whirled rapidly in the air for about a minute. The moisture from the wick evaporates into the air due to the movement of air across it, and this cools the wet bulb. The drier the air, the more water evaporates, and the lower the wet-bulb reading. By placing the wet-bulb temperature and the dry-bulb temperature readings opposite one another on the slide included with the psychrometer, relative humidity can be read directly as a percent. Or, the wet-bulb depression chart can be used to calculate relative humidity. Digital psychrometers are also available.

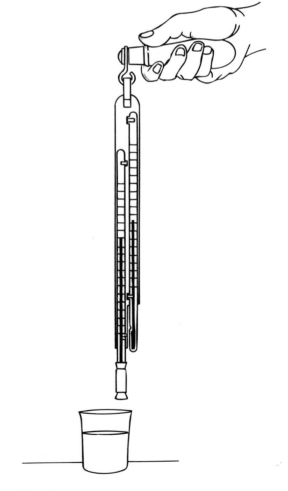

Figure 12-2. Sling Psychrometer

DEW-POINT TEMPERATURE

Dew-point temperature is the temperature at which water vapor begins condensing out of the air. For example, when air is at 70 °F and 50 percent relative humidity, its dew-point temperature is 51 °F. If this air comes into contact with a window whose surface temperature is 50 °F, the water in the air will condense onto the window. If the window surface temperature is considerably lower than the room temperature (i.e., 32 °F), the moisture on the window will freeze, forming ice and frost on the window. This is not desirable, because when the frost melts, it can cause structural damage by dripping onto the window sills and frames.

When frost or condensate form on a window, it is a sign to the service technician that the humidity level in the living space is too high. This humidity level can be adjusted by the humidifier (to be discussed shortly). The type of glass in the window also determines how fast frost or condensate will form on a window. For example, moist air condenses on single-pane glass at much higher outside temperatures than on double-pane, or insulated, glass, Figure 12-3. Thus, when insulated glass is in place, a higher relative humidity may be maintained within the living space without condensation.

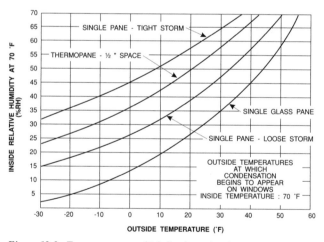

Figure 12-3. *Temperatures at which Condensation Appears on Window Glass*

It is necessary to know the dew-point temperature when dehumidifying an area. This is because dehumidifying is accomplished by passing air over a surface whose temperature is below the dew-point temperature of the air. When this occurs, the moisture from the air collects on the cold surface and is then drained. For example, when using a window air conditioner, this moisture collects in the evaporator and then drains out the back.

INFILTRATION

Infiltration is the process of warm, moist indoor air escaping outside (through doors, windows, etc.) and being replaced by cold, dry, outside air. Infiltration occurs no matter how well-constructed the building may be; however, several factors, such as outside high wind velocities or cold outside air temperature, can speed up the process.

The amount of cold air entering a structure is measured in air changes per hour. An air change is the quantity of outside air that enters and displaces all of the inside air in the total cubic contents of the structure. The number of air changes that occur in an hour depends upon the construction of the house:

- If a house is tight, with insulated walls and ceiling, tight-fitting storm doors and weather-stripping, air infiltration is about half an air change per hour.
- If the house structure is average, with insulation in the walls and ceiling, loose-fitting storm windows and doors, air infiltration is about one air change per hour.
- If the structure is loose, with no insulation, no weather-stripping, and no storm doors or windows, air infiltration is about two air changes per hour.

If large volumes of cold, dry outside air are constantly coming into a structure and circulating through the heating system, the relative humidity is very low. This is because when outdoor air at 0 °F and 90 percent relative humidity is heated to 72 °F indoor temperature, without adding moisture, its new relative humidity is only 5 percent. For this reason, a humidifier is needed.

HUMIDIFIERS

Many factors add moisture to the living space, including activities such as cooking, showering, and dishwashing. However, these activities are usually not substantial enough to offset the very low relative humidity of the heated outside air. Also, these moisture-producing activities are usually concentrated at certain times of the day, so they do not provide a continual source of moisture.

The only satisfactory solution to humidity control is installing a humidifier whose capacity is matched to the particular requirements of the residence. A humidifier is integrated into the heating system and introduces additional moisture into the duct system. Some units introduce a fixed amount of water per day or per hour, while others are controlled by a humidistat, which monitors the moisture content in the house and turns the humidifier on or off to maintain a preset relative humidity.

A humidistat senses the amount of humidity in the air and turns the humidifier on and off according to design requirements. It can be located on the wall in the living room, next to the thermostat, or it can be located in the return air duct. The former location is more accurate, but the latter is easier because it involves less complicated wiring. Use of a humidistat is preferred, because it holds the relative humidity at a comfortable, preset level. This is important, because it is not desirable to raise the relative humidity too high during the winter.

There are basically four types of humidifiers: pan, atomizing, wetted-element, and infrared. All four introduce water vapor into the house air stream at predetermined rates, and all require a source of water. Most of them require a drain connection, and some require a humidistat, with additional field wiring, and a blower or rotor motor. All designs, however, follow the same basic principles.

PAN-TYPE HUMIDIFIER

The earliest type of humidifier is the pan-type humidifier. The pan-type humidifier uses heat from the air or from an external source to vaporize the water. The vaporized particles then mix with the heated air supply. This type of humidifier is also called an evaporative humidifier, because it works on the principle of passing dry air over moisture on a particular surface. The dry air then picks up this moisture and carries it to the living space.

A pan-type humidifier is constructed of a rack resting in a pan of water, Figure 12-4. Porous, mineral, wool plates are then inserted into the rack (this plate design increases the surface area and the humidifying capacity). The bottoms of the plates rest in the pan water and absorb water by wick action until the plates are completely wet. The pan, with the plates, is inserted into the furnace plenum where the warm supply air picks up moisture from the plates as it passes over them. The water level in the pan is maintained by a float valve.

To increase the capacity of a pan-type humidifier, a steam coil or electric heater may be immersed in the pan of water. This helps increase the water temperature, thus increasing the rate of evaporation. When the water is free of contaminants, this combination of a high temperature element and water is effective. However, if the water used contains a high mineral content, the higher temperature may cause scaling, which eventually burns out the element.

This type of humidifier is installed in either the furnace supply or return-air plenum. The water supply and drain are then connected. The furnace fan blows over the humidifier, distributing humidity to the living space. There is no electrical requirement as the float valve maintains a preset, constant water level. This type of humidifier has a limited capacity and works well in smaller homes.

When wired correctly, humidifiers requiring the furnace blower operate regardless of whether the burners are on or off. For best results, the furnace blower should be kept on at all times in order to move the greatest amount of air over the humidifier. Furnaces from many manufacturers have a terminal board in the furnace to make either 24-V or 115-V humidifier connections.

ATOMIZING HUMIDIFIERS

An atomizing humidifier converts water into very small droplets before introducing them into the air stream. This can be accomplished in two ways: by spinning disc or high-pressure spray nozzle.

The spinning disc method involves a circular wheel or cone that rotates at a fairly high speed, Figure 12-5. Water is fed onto the rotating wheel and centrifugal force converts it into small droplets. This method is often used with self-contained humidifiers.

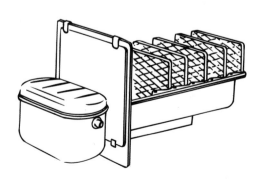

Figure 12-4. Pan-Type Humidifier

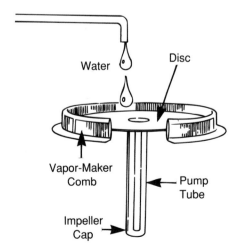

Figure 12-5. Spinning Disc Atomizing Humidifier

The high-pressure spray nozzle blows water onto a pad or splash plate, Figure 12-6. This type of humidifier can attach to the supply air duct with a bypass to the return air duct. This humidifier works due to the air pressure differential between the supply and return sides of the system. This pressure differential draws air through the humidifier, thus picking up moisture from the pad or splash plate. The air then moves to the supply duct, and then circulates throughout the living space.

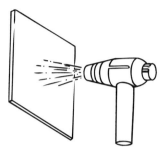

Figure 12-6. *High-Pressure Spray Nozzle Atomizing Humidifier*

The atomizing humidifier can cause several problems. The first problem is hard water. Hard water, due to its high mineral content, should not be used in an atomizing humidifier. These minerals leave the water vapor as dust and subsequently settle throughout the living space. The second problem occurs when water particles do not vaporize completely. These droplets settle in the air distribution system and can cause rust or corrosion.

WETTED-ELEMENT HUMIDIFIERS
Wetted-element humidifiers are possibly the most common type used in forced warm air systems. These humidifiers use porous media which is wetted by water. Air passes over the media by either an air pressure differential or a self-contained fan. This air current picks up moisture from the media and carries it into the house duct system. It is important to keep in mind that among all systems working on air pressure differential, the air always flows from the warm air plenum to the return air plenum, regardless of where the unit is mounted.

Rotating Drum. The rotating drum is one type of wetted-element humidifier, Figure 12-7. This unit consists of a drum covered with a screen or sponge pad. This drum rotates, due to a 24-V motor, very slowly (1 rpm) through a pan of water. A float valve, included with the assembly, keeps the water in the pan at a constant level (about 1 3/8 inches). The bottom part of the pad is immersed in the water and the entire pad becomes wet as the drum rotates. Air is drawn across the wet pad due to the air pressure

difference between the supply and return sides of the system. The air picks up moisture and returns to the house duct system.

Figure 12-7. *Rotating Drum. Courtesy, Skuttle Manufacturing Company.*

A humidistat wired in series with the 24-V drum motor controls the operation of a drum. The humidistat can either mount on the wall, near the thermostat, or in the return air duct. If mounted in the duct, care should be exercised to be sure it is out of the path of any radiant heat. Figure 12-8 shows the wiring for this unit. With this arrangement, the humidifier operates independently of the furnace cycle and provides additional moisture whenever the humidistat senses a drop in humidity.

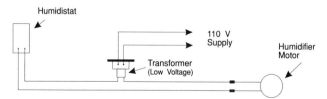

Figure 12-8. *Humidistat Wired in Series with Humidifier Motor*

Disc Screens. Another type of wetted-element humidifier uses disc screens. These screens are each approximately 10 inches in diameter and rotate in a plastic reservoir of water which mounts horizontally under the supply air duct, Figure 12-9. When the discs rotate, warm air blows across them, picking up the water in the screens and carrying the humidified warm air to the living space. All water impurities are left behind on the screens, so the water vapor is pure. However, these discs occasionally need to be cleaned in order to rid the screens of impurity build-up.

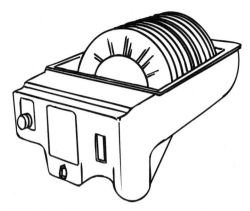

Figure 12-9. Wetted-Element Humidifier Using Disc Screens

Top-Feed System. A top-feed system is another way to utilize the wetted-element principle. This type of humidifier uses a porous material which is mounted vertically. Water from the top then drips onto the material and moves evenly across the entire top surface. The porous material must allow the water to pass slowly from top to bottom to maintain a uniformly wet surface. Air is then forced through the wetted medium, Figure 12-10.

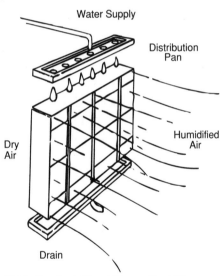

Figure 12-10. Top-Feed, Wetted-Element Humidifier

In a top-feed system, an arrow on the porous pad indicates the proper direction of air flow. This type of unit also requires a drain connection, since the water that does not evaporate drips down to the bottom of the unit. Also, as there is no reservoir of water, the water flow must be controlled by a humidistat and allowed to flow only when the humidistat senses a drop in humidity. This is accomplished by wiring a solenoid water valve in series with the humidistat. When the humidistat contacts close, they energize the solenoid coil, opening the valve and allowing the water to flow into the top of the unit. Figure 12-11 shows the basic wiring for a 24-V system. This wiring allows the humidifier to operate independently of the furnace cycle.

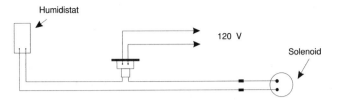

Figure 12-11. Basic Wiring in 24-V System

This unit mounts directly on the supply side of the plenum and has a bypass pipe running from the face of the unit to the return air side of the system, Figure 12-12. Warm air is drawn through the wetted pad by the air pressure differential between the two sides of the system. The moving air picks up moisture, by evaporation, and returns to the duct system.

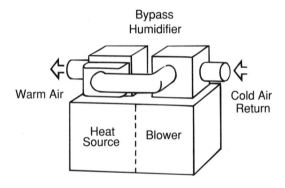

Figure 12-12. Vertical, Wetted-Element Humidifier Installed with Bypass Pipe

Another variation of this top-feed system has the pad in the vertical position with water feeding it through a solenoid operated water valve as before. However, the air circulates through the porous pad due to a separate fan built into the humidifier. In this configuration, there is no need for a bypass pipe, and the unit mounts directly on the furnace plenum. Water supply and drain lines are required, as in the previous version. Air bypass openings flank the pad on both sides, allowing the fan to draw warm air directly from the plenum and then push it through the evaporation media and back into the plenum, Figure 12-13.

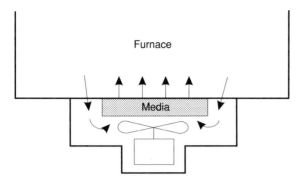

Figure 12-13. Top-Feed, Fan-Equipped, Wetted-Element Humidifier

Location. Wetted-element humidifiers can be mounted in various positions, according to the design. Some units fit under the supply duct, or on the furnace plenum itself. Other units require a fan which draws air from the furnace plenum, through the humidifier, then returns the air back to the plenum. This type of unit mounts on the supply duct, and the furnace blower circulates the warm air through the humidifier and supply duct, and finally, to the return air duct. The manufacturer of the humidifier provides explicit instructions on where the humidifier should be placed.

Wiring. Fan motors and solenoid water valves found in the wetted-element humidifiers are 115 V. A line or low voltage humidistat controls the motors and valves. Figure 12-14 shows the humidistat in the line voltage circuit. Figure 12-15 shows the humidistat in the low voltage circuit. When wired in the low voltage circuit, a relay must also be included.

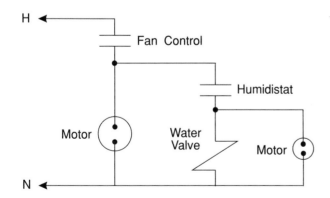

Figure 12-14. Humidistat Wired Into Line Voltage Circuit

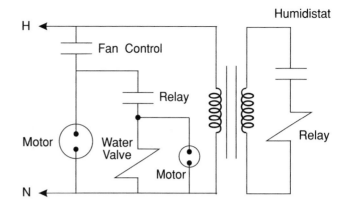

Figure 12-15. Humidistat Wired Into Low Voltage Circuit

INFRARED HUMIDIFIERS

This type of humidifier contains infrared lamps which reflect infrared energy onto water in a horizontal water reservoir. Water vapor does not form due to boiling; instead, it rises off the water surface at temperatures rarely exceeding 170 °F. Infrared radiation is a large part of solar energy, and this radiation allows the sun to vaporize water from lakes, rivers, etc. This type of humidifier operates on the same principles as the sun.

The infrared humidifier does not require heat energy from a heating system to vaporize water, as the lamps provide their own heat. Therefore, infrared humidifiers can be installed in low temperature locations. Some of these low temperature locations include downflow, electric, and horizontal furnaces.

HUMIDIFIER SIZING

Humidifiers are usually rated in terms of gallons of water they deliver per hour or per day into the house duct system. The total gallons required for an installation depends upon the cubic capacity of the house, its construction, the outside winter temperature, the desired inside relative humidity, and the heating system temperature.

Most manufacturers provide a set of guidelines to help the service technician determine the proper size and type of humidifier for a given house, using the variables discussed previously. For example, one manufacturer suggests that a 1 gallon/hour (gph) humidifier can handle a loosely-constructed 11,000 ft³ house, an average 22,000 ft³ house, or a tightly-constructed 44,000 ft³ house. The cubic capacity of a house is calculated by multiplying the number of heated square feet in a house by its ceiling height. For example, a 2,000 ft² house with a ceiling height of 8 feet has 16,000 ft³ (2,000 ft² x 8 feet).

HUMIDIFIER MAINTENANCE

As stated previously, all water contains a certain amount of minerals, excepting distilled water or rainwater before it reaches the ground. When water evaporates, these minerals in the water remain as dust particles which can cause problems.

Water is classified by its hardness, or the number of grains of mineral matter it contains per gallon. This varies considerably from area to area, and even varies within the same city. There are three classes of hardness: low, average, and high. The low class contains 3 to 10 grains/gallon, the average class contains 10 to 25 grains/gallon, and the high class contains 25 to 50 grains/gallon. Approximately 30 percent of the people in the U.S. use low water hardness, 55 percent use average water hardness, and 15 percent use water with a high hardness level.

The harder the water in a given area, the greater the problem with mineral deposits in the humidifier. In some areas with very hard water, no type of humidifier functions properly. Humidifier maintenance consists primarily of preventing serious build-up of these mineral deposits.

The first and most important maintenance procedure is to thoroughly clean all operating parts. A 50 percent vinegar and water solution is a good solvent for most mineral deposits found on the operating parts. It is also necessary to change media pads regularly. Polyurethane media can be removed, washed in soap and water, and replaced. Extreme water hardness may require changing pads and cleaning the system every 4 to 6 weeks.

A plastic enclosure, or some other kind of enclosure, protects most motors used in humidifiers. These enclosures help reduce the accumulation of mineral deposits on operating parts. The motor needs oiling every season. Electrical connections, such as solenoid valves, are vulnerable to mineral build-up, even though they are protected by plastic sleeves. Float valves can also acquire a mineral coating and if they do, they either function erratically or not at all. Even models with regular flushing cycles, to minimize mineral buildup, can end up with clogged drains. Electrical checks should also be made on the transformer, humidistat, fan motor, solenoid, and drum motor to determine if each is operating properly (not all of these parts apply to every humidifier).

Spray nozzles are particularly vulnerable to mineral build-up, and it is a good idea to alternate nozzles every year. The nozzle not used should then be soaked in vinegar until the next season. Mineral build-up at the needle valve in the main supply can result in water being cut off to the unit. This valve should be checked each year.

REVIEW QUESTIONS

1. What is the purpose of a humidifier? COMFORT, FURNISHINGS
2. What is the unit of measurement for moisture content of air? gr./lb.
3. Define relative humidity.
4. What happens to relative humidity when air temperature increases? ↓
5. What is infiltration? LEAKS
6. How is house construction defined? TIGHT, LOOSE, AVERAGE
7. What is the normal relative humidity level? 40%-60%
8. How are wet-bulb and dry-bulb temperatures calculated?
9. What is wet-bulb depression? DRY - WET
10. What is dew-point temperature? CONDENSE
11. Why is condensation a problem? MILDEW, WATER DAMAGE
12. What does a humidistat do? MEASURES HUMIDITY
13. Name the four basic types of humidifiers. DRUM PAN, ATOMIZE, INFRA, WETTE
14. How does a pan-type humidifier work? EVAPORATON
15. Describe the high-pressure spray nozzle method. - ATOMIZER
16. What are the major problems of an atomizing humidifier? RUST + SCALE
17. Name three different wetted-element humidifiers. What are their differences? DRUM, DISC SCREEN, TOP FEED
18. Name several possible locations for an infrared humidifier. DOWNFLO, ELECTRIC OR UPFL HOR.
19. How does a service technician select a humidifier for a house? BY SIZE + CONSTRUCTION TIGHTNESS
20. What is the importance of water hardness? SCALE

Chapter 13

Air Cleaners

Polluted air contains unacceptable concentrations of foreign substances over and above its normal makeup of 21 percent oxygen, 78 percent nitrogen, 0.03 percent carbon dioxide, and minute traces of other elements. Dirt, dust, and ash are pollutants. These pollutants permeate the air when events such as earthquakes, windstorms and forest fires occur. Industrial plants, incinerators, automobiles, and the like, also create pollutants. Pollutants can even evaporate into the air from polluted water.

Pollutants may occur as gases, oxides, hydrocarbons, particles and vapors. Some types of pollutants are merely irritating, while others, especially pollen, viruses and bacteria, can move into the lungs or circulatory system and cause colds, allergies and other serious health problems. These pollutants circulate through every building. Consequently, air cleaners are necessary to help remove polluting or irritating particles from a living space.

MEASUREMENT OF POLLUTANTS

Particle pollution makes up 13 percent of total air pollution. These particles may be either solid or liquid, and they act as carriers of dangerous contaminants, such as sulfur dioxide. It is possible to control these particles through electronic air cleaning.

A particle is an object which has definite physical boundaries in all directions, and has a diameter ranging from 0.001 microns to 100 microns. A micron is 0.000001 of a meter. It takes 25,400 microns to equal one inch, or about 200 microns to reach across the dot made by a sharp pencil. A particle of fine sand measures about 100 microns.

The visible particles are 10 microns and larger and make up only about 10 percent of the total number of airborne particles; the other 90 percent are invisible. Particles measuring less than 1 micron remain suspended in the air indefinitely. However, oily or greasy particles (0.01 to 1.0 microns) eventually come in contact with a vertical or horizontal surface and adhere to it.

MECHANICAL FILTER SYSTEMS

All furnaces are equipped with some sort of device for filtering air. As stated in Chapter 4, the filtering device usually contains a medium of Fiberglass, polyurethane or metal. The medium is oiled or coated on one side to collect and hold airborne particles as they pass through the filter. This filter is a mechanical filter and eliminates particles down to about 10 microns. This includes dust, lint, pollen, plant spores and fly ash. Smaller particles normally pass freely through the filter and are recirculated through the house. The mechanical filter only removes approximately 10 percent of the particles in the air, as opposed to an electronic air cleaner which removes about 95 percent.

As the mechanical filter collects dust and lint, its cells become more constricted. When this occurs, smaller particles also become trapped, due to the dust and lint blocking their path in the filter. This condition creates a greater resistance to air flow, reducing efficiency. For this reason, filters must be changed or cleaned periodically in order to maintain proper temperature and air circulation.

ELECTRONIC AIR CLEANERS

Electronic air cleaners (EAC) are designed to eliminate smaller particles, measuring from 10 microns down to 0.01 microns, from the air. These particles include viruses, tobacco and oil smoke, and similar particles. A good EAC eliminates 95 percent of the particles in this size range, Figure 13-1.

Figure 13-1. Electronic Air Cleaner. Courtesy Carrier Corporation, a Subsidiary of United Technologies Corporation.

OPERATION

The process by which the EAC traps particles is called electrostatic precipitation. An EAC is actually a two-stage electrostatic precipitator. The first stage is an ionizing section, and the second stage is a collecting section, Figure 13-2. Each of these stages is housed in a section called a cell, and together they make up the electronic, or main, cell.

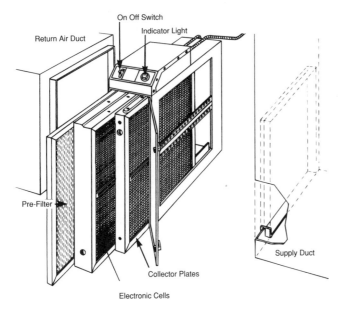

Figure 13-2. Electronic Air Cleaner, Disassembled

The ionizing section consists of very small-diameter tungsten wires. These wires connect to the 8000-vdc output of the power pack. This high voltage creates a positively-charged electrostatic field around each wire.

As the particles in the air move toward this field, electrons are knocked off atoms (atoms make up molecules, which in turn, make up particles). These electrons are called free electrons and can strike other air molecules and knock more electrons out of them. The molecules are left with a surplus of positive charge and are called positive ions. These ions then bombard the airborne particles. In this manner, particles become either positively or negatively charged. Once these charges are established, the particles move toward the collector plates in the collecting section.

The collecting section also contains an electrical field, however, this field is set up between a series of grounded collector plates made of metal or screen material. These collector plates carry positive and negative charges. The positively-charged particles move toward the negative plates, while the negatively-charged particles move toward the positive collector plates. These particles remain on the plates until they are washed off. Figure 13-3 shows the electronic cell.

Figure 13-3. View of Electronic Cell. Courtesy Carrier Corporation, a Subsidiary of United Technologies Corporation.

As the particles collect on the plates, they lose their charge and assume the charge of the plate. These particles build up on the plate, insulating and reducing the efficiency of the plate. Therefore, it is necessary clean the plates periodically to maintain proper operation.

An EAC operates at a high degree of efficiency when cleaning air, and it does not restrict air flow. The actual pressure drop across an EAC is no greater than that for a common throwaway mechanical filter. The efficiency of an EAC does increase, however, when air velocity is lower; therefore, the EAC should be sized as closely as possible to the actual amount of air circulated in the house.

TYPES OF EACS

There are several different types of EACs, depending on the application. There are self-contained models, which

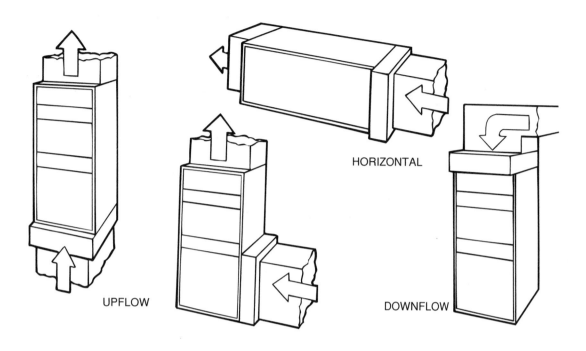

Figure 13-4. Electronic Air Cleaner Installations. Courtesy Carrier Corporation, a Subsidiary of United Technologies Corporation.

are easily movable and can be used to clean air in a specific area. These models contain a built-in, motor-driven fan which helps circulate air throughout the unit. These units are normally employed in locations which do not have a central heating system. The other type of EAC is permanent, and mounts in the duct system of a central heating or air conditioning system. The air moves through this type of EAC by way of the fan in the system.

INSTALLATION

The typical EAC operates in the return air duct of a forced air system, ahead of the furnace blower. This allows the EAC to remove more foreign particles each time the return air passes through the unit. If a prefilter is used, it is installed ahead of the EAC, and the regular furnace filter is removed.

Depending on the unit, the ionizer and collecting plate cells can either be removed separately, or else the entire electronic cell can be removed. The EAC can be located almost anywhere in the furnace return air system, Figure 13-4. It is necessary to leave room in the front of the unit to slide the cell in or out of the cabinet.

The system may also require a prefilter made of aluminum mesh, used to trap the larger particles measuring 10 microns or more. Figure 13-5 shows the entire process of an EAC.

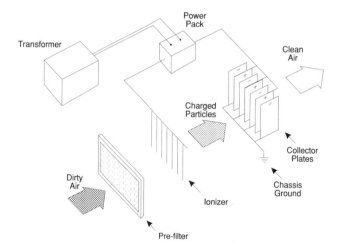

Figure 13-5. Electronic Air Cleaner Process

If a fresh air intake is used, it must be located upstream of the EAC. This intake must be screened and arranged so the outside elements (i.e., snow, rain), as well as insects and leaves, do not interfere with the EAC. In order to prevent cold air from reaching the EAC, preheat coils are needed in this situation. The return air temperature across the EAC should not drop below 40 °F.

When a humidifier is used in conjunction with an EAC, care should be taken to make sure no water from the humidifier reaches the EAC. As water is a conductor, it causes popping and cracking, due to arcs between the plates. Also, the minerals in the water cause the plates to deteriorate, which reduces the life of the EAC.

ELECTRICAL COMPONENTS

While all EACs have essentially the same components, the wiring required and the voltages produced may vary. It should also be noted that all air cleaners incorporate both alternating current (ac) and direct current (dc). For this reason, it is necessary to check the manufacturer's diagram carefully to determine the type of current in each circuit of a given unit.

The basic EAC components and their functions are as follows (the last three components listed are available only on certain EACs):

Pilot Light. Some EACs contain a pilot light wired into the transformer's resonator circuit. This light remains on whenever the unit is functioning properly. It turns off when power is disconnected, the door is opened, service is needed, and when the plates need cleaning.

Ammeter. Certain models use an ammeter to visually indicate when the equipment is functioning properly and when maintenance or service are needed. The ammeter is wired into the line (120 vac) voltage circuit.

Off-On Switch. An off-on switch is required, so the EAC section can be de-energized during the drying cycle.

Power Disconnect or Interlock. The interlock is a provides a means of disconnecting power to the EAC whenever it is necessary to remove the cell for cleaning or service. This safety feature prevents a possible shock hazard.

Step-Up Transformer. The transformer converts the 120-vac primary to a much higher voltage (approximately 4000 vac) in the secondary. In order to be effective, an EAC requires this very high voltage.

Rectifier Circuit. The rectifier circuit consists of two selenium rectifiers (diodes) and two capacitors. (In some EAC models, the collector plate acts as one capacitor.) The purpose of the rectifier is to convert the ac voltage input to dc voltage for the collector and ionizer. The capacitors are used in a voltage double circuit, which approximately doubles the voltage in one leg of the circuit.

Bleeder Resistor. This resistor bleeds off to ground any excess voltage remaining when the unit is de-energized. Some units perform this function by causing the interlock switch to short across the plates when the door is opened. This eliminates any shock hazard.

Collector Plates. Collector plates are spaced 3/16 to 1/4 inch apart. Every other plate is connected to the positive side of the rectifier circuit, and the remaining plates are connected to the ground.

Ionizer Wires. These small-diameter wires are mounted with tension springs and insulators that are connected to the positive side of the rectifier circuit.

Built-In Water Wash. This component allows the collector plates to be washed in place. When a system contains this component, a water supply and drain connection are required.

Dry-Cycle Timer. The dry-cycle timer has a clock motor which can be set from 1 to 3 hours. The timer de-energizes the EAC circuit to allow adequate drying time after the cells have been washed.

Blower On Switch. This switch ensures that the blower stays on and does not cycle with the fan control during the drying cycle.

FIELD WIRING

EACs are completely prewired at the factory. The only field wiring required is to connect a 120-V power supply from the furnace electrical system either directly into the EAC or to a makeup box on the EAC. Some furnaces have a terminal block with provision for EAC connections.

For efficient operation, the EAC should operate only when the furnace blower is on. There are two ways to accomplish this. One method is to wire the EAC into the fan control circuit so it only runs when the fan control turns on the blower. This method usually is the most convenient with a single speed motor. The second method is used with a two-speed blower. In this case, the EAC on time is controlled by a sail switch, which is mounted in the return air duct. This switch has a plate, or sail, which extends into the air stream and mechanically connects to a normally-open switch. When the air pressure in the return air duct increases (blower on), it moves the sail and closes the contact, turning on the EAC, Figure 13-6.

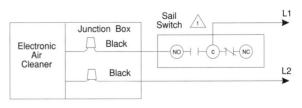

⚠ Sail switch mounted in return duct
   energizes EAC when fan is running.

Figure 13-6. Schematic Wiring for Sail Switch

INTERNAL WIRING

The internal wiring and basic components of the power pack are very similar for all EACs. The ionizer voltage is around 8000 vdc, and the collector voltage about 4000 vdc, with respect to ground. A schematic wiring diagram for one unit is shown in Figure 13-7. All EACs have a step-up transformer to change the 120-vac input into the secondary voltage of approximately 4000 vac.

MAINTENANCE

Basic maintenance of an EAC entails the removal and cleaning of the electronic cell at regular intervals. This cell normally slide out easily and can then be washed. There are models available that even allow the service technician to wash the cell without removing it. Even better, there are EACs available with a light bulb system that indicates when the electronic cell needs to be cleaned.

The frequency of cleaning depends upon a number of variables. One variable is the construction of the house, as discussed in Chapter 12. Loose construction allows more particles to enter the house than tight construction. And, if the residents smoke or entertain frequently, an air cleaner must work even harder, and the dirt buildup on the plates accelerates.

Normally, the electronic cell should be cleaned every two to three months, and more often if unusual conditions exist or if the plates become exceedingly dirty between cleanings. In addition, the cells should be cleaned any time the pilot light is out or the performance meter is in the service position.

To clean the electronic cell, perform the following steps:

1. Remove the electronic cell.
2. Place the cell in an automatic dishwasher with the ionizer side down. Make sure the air flow arrows are pointing up.
3. Add detergent and run the dishwasher through the complete washing and drying cycle.
4. Allow the cell to cool after the drying cycle, to avoid damaging the plates or breaking one of the ionizer wires.
5. Drain any accumulated water from the tubes supporting the collector plates.

If a dishwasher is not available, perform the following steps:

1. Place the cell in any container large enough to hold it, such as a laundry tub or sink.
2. Lay the cell flat in the tube and completely immerse it in dishwashing detergent mixed with hot water.
3. Soak the cell for about 15 to 20 minutes.
4. Rinse the cell with a fine spray and then soak it in clean, hot water for 5 to 15 minutes.
5. Remove the cell and let it drain. The cell can then be reinstalled.

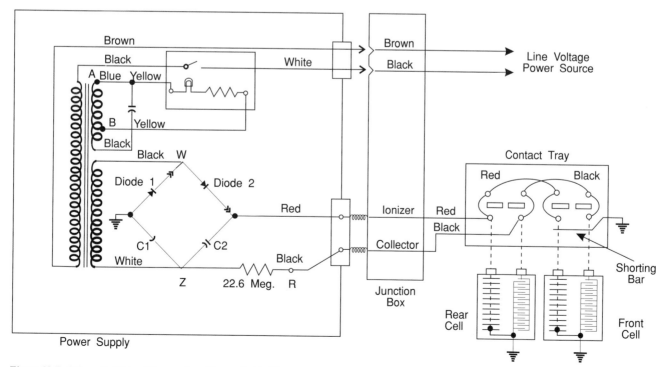

Figure 13-7. Schematic Wiring Diagram from Electronic Air Cleaner

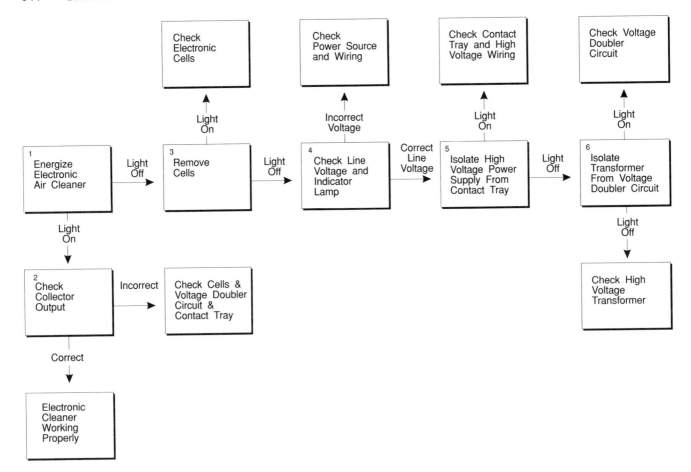

Figure 13-8. *Troubleshooting Flow Chart*

After the clean cell is reinstalled in the unit, it must be thoroughly dried. This is accomplished by running the furnace blower for at least an hour with the EAC disconnected. Units with a dry-cycle timer automatically disconnect the cell until the drying cycle is completed, and then re-energize the EAC power pack. The manual switch turns off units without a dry-cycle timer. The units must be turned on again once the drying cycle is finished. Also, the fan control must be bypassed either by a switch on the unit or by setting the thermostat for continuous blower operation.

Any time the cell is removed from the unit for cleaning, perform the following inspections:

- Inspect the cell for any particles lodged in the plates, as these could cause a short.
- Check for bent or shifted plates. Straighten any bent or shifted plates.
- Inspect ionizing section for cracked or broken insulators.
- Inspect and replace broken wires in ionizing cell. (The collecting cell functions, but at reduced efficiency, when a broken wire is removed and not replaced.)

SERVICE

Servicing an EAC is not difficult, as most problems occur due to dirty or incompletely dried cells. The pilot light or performance meter indicates when the unit requires attention. If washing and drying the cell does not correct the problem, additional checks must be made. The following tools are then required:

- Two screwdrivers with insulated handles
- Needlenose pliers
- Test meter with a range of at least 12,000 vdc
- Soldering iron
- Neon test light (120 V)

The troubleshooting steps for most EACs can be performed by observing the indicator light in the on/off switch. The light is on whenever the transformer is working properly. A flow chart for electrically troubleshooting an EAC is shown in Figure 13-8. In other EACs, it is necessary to substitute ammeter readings or voltage readings for pilot light operation.

As stated earlier, because of variations in voltages and test points, the service procedures suggested by the manufacturer of the EAC being serviced should be followed. Some general suggestions, following the sequence given in the chart, are as follows:

- Be sure cells are clean and dry, and there are no broken wires or shorts.
- Check collector output by shorting across two plates. An arc or snapping sound indicates the EAC is working properly.
- Check the line voltage supply with a meter or test light to verify a 120-V supply. If there is no supply, check backwards in the circuit through the switch to the power source.
- Isolate the high voltage supply, disconnect the cells and check the output voltage. If correct, check the high voltage wiring. If incorrect, isolate the transformer and check the output voltage.
- If the transformer output voltage is correct, check the rectifier circuit. If not correct, check the transformer windings for continuity.
- Check the voltage across each capacitor in the rectifier circuit. If low, check the rectifiers. A selenium rectifier can be checked by pushing on the end at which the arrow points, with a pencil. If the spring does not compress, the rectifier is bad and needs replacement.
- Capacitors can be checked, out of the circuit, with a meter. A good capacitor moves part way on the scale and then to infinity. A shorted capacitor shows zero resistance.
- On units with a bleeder resistor, turn on the unit for a few minutes, then turn it off. Wait about 10 seconds and then short across two adjacent collector plates with an insulated screwdriver. An arc indicates that the resistor is bad.

REVIEW QUESTIONS

1. What are pollutants?
2. What causes pollutants?
3. What is a particle?
4. How big are visible particles?
5. What size particles does a mechanical filter remove?
6. What does an activated charcoal filter do?
7. How does an EAC work?
8. What size particles does an EAC remove?
9. Where is an EAC located?
10. How is an EAC wired into the furnace?
11. What special precautions must be taken when an EAC is installed with a humidifier?
12. Describe three electrical components in an EAC.
13. How can you tell when the cell needs cleaning?
14. How often should the cell be cleaned?
15. Name four things to check for in regular maintenance.

Chapter 14

Heat Pumps

A heat pump is similar to an air conditioning system, except the heat pump moves heat to and from the living space, Figure 14-1. When properly employed, the heat pump can be used all year. In the summer, it acts as an air conditioner. In the winter, the heat pump reverses the refrigerant flow and becomes a heating unit. Because of its ability to reverse refrigerant flow, the heat pump is also called a reverse cycle machine.

Figure 14-1. Heat Pump. Courtesy, Carrier Corporation, a Subsidiary of United Technologies Corporation.

The heat pump is a practical system for residential heating and cooling for several reasons: (1) there are no vents or chimneys required, as there are no products of combustion, (2) the equipment takes up a small area, and (3) the components are effective, efficient, and dependable.

There are three basic types of heat pump systems: air-to-air, water-to-air, and ground-to-air. Also, some heat pumps, like air conditioning systems, come in either packaged systems, having all the components in one cabinet, or split systems, with the outdoor unit remote from the indoor unit. While all of these heat pumps are discussed in this chapter, all examples of heat pumps (i.e., detailing the heating and cooling cycles) use the air-to-air, split system heat pump design.

HEAT PUMP OPERATION

A heat pump is not simply an air conditioner with the capability of reversing refrigerant flow. The heat pump is a precise, carefully engineered, matched-component system which can operate in three different modes: cooling, heating, and defrosting. In a split system, the outside unit looks exactly like a conventional air conditioning condenser. The inside unit is somewhat smaller than a furnace.

Before exploring the three heat pump modes, it is necessary to understand the two components found in a heat pump system, which are not found in a refrigeration system: heat pump coils and four-way valves.

HEAT PUMP COILS

The difference between heat pumps and refrigeration units is that heat pumps do not have evaporators or condensers. Instead, heat pumps use coils, which function as evaporators or condensers. The heat pump "evaporator" is called the indoor coil, and the "condenser" is called the outdoor coil. The function of each coil changes, depending on whether the unit is in the heating or cooling mode. For example, when in the cooling mode, the indoor coil performs the job of an evaporator by gathering heat from the living space. In the heating mode, the indoor coil performs the job of a condenser by rejecting heat into the living space. By reversing the refrigerant flow, the function of the coils change. The terms indoor and outdoor are necessary to distinguish whether the coil is acting as an evaporator or a condenser.

FOUR-WAY VALVE

This valve is also called a reversing valve, Figure 14-2. This component allows the refrigerant to flow in two directions, so the indoor coil can receive either hot or cold refrigerant. The four-way valve is housed between the compressor and the indoor and outdoor coils. By changing flow direction inside the body of the four-way valve, the refrigerant flow actually reverses in the system.

Figure 14-2. Four-Way Valve. Courtesy, Ranco North America.

There are four tubes which branch off the four-way valve. A single tube connects to the discharge side of the compressor. On the opposite side of the valve are three tubes. The two outside tubes connect to the indoor and outdoor coils, and the center tube connects to the compressor suction line, Figure 14-3.

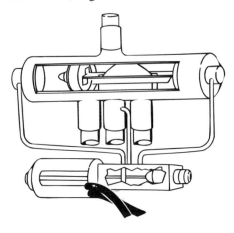

Figure 14-3. Four Tubes in Four-Way Valve

The four-way valve main body encloses a slide assembly, which consists of two pistons, one mounted on either side of a sliding block. High-pressure compressor gas causes the slide assembly to move from one end of the four-way valve to the other. This action causes the slide assembly to cover and uncover the tubes leading to the indoor and outdoor coils, determining the direction of refrigerant flow. When the unit is in the heating mode, the indoor coil receives the discharge gas from the compressor. In the cooling or defrost mode, the outdoor coil receives the compressor discharge gas. This demonstrates that the coil ports are alternately suction or discharge ports, while the gas flows through the compressor in the same direction at

all times. The four-way valve main body also contains a small orifice on either end of the slide assembly. These orifices allow discharge gas to enter the valve main body and accumulate between the slide and the four-way valve body wall.

A solenoid-operated pilot valve determines the position of the slide in the main valve body. The pilot valve has an internal plunger which, when energized, extends or retracts, opening a suction line on one side of the four-way valve body. Discharge gas on one side of the slide bleeds into this suction line, while discharge gas builds up on the other side of the slide. This causes an imbalance of pressure in the valve, with low pressure on one side and high pressure on the other. The slide moves toward the low pressure side. This slide movement causes the discharge gas to flow to the alternate coil. When this occurs, refrigerant flow changes direction, which switches the unit mode of operation. In the heating mode, the plunger is retracted, Figure 14-4, and in the cooling mode, the plunger is extended, Figure 14-5.

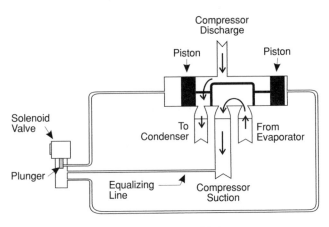

Figure 14-4. Pilot-Valve Plunger in Heating Mode

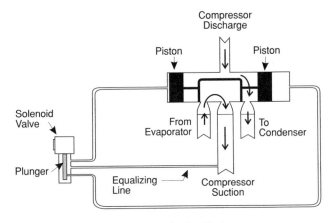

Figure 14-5. Pilot-Valve Plunger in Cooling Mode

COOLING MODE

In the cooling mode, the indoor coil acts as an evaporator, absorbing heat from the living space. This heat causes the liquid refrigerant in the "evaporator" to boil, and the refrigerant turns to a low-pressure gas. The gas passes through the four-way valve and moves to the accumulator in the suction line (the line between the four-way valve and the compressor). As in a regular air conditioner, the gas then travels to the low-pressure side of the compressor. The compressor takes the low-pressure gas and compresses it into a high-pressure gas and then discharges the gas through the discharge line (the line between the compressor and the four-way valve). The high-pressure gas moves through the four-way valve again, and then enters the outdoor coil, which acts as a condenser. (The refrigerant bypasses the metering device at the outdoor coil.) The "condenser" cools the high-pressure gas into a high-pressure liquid, and then this liquid moves through the strainer and drier. The liquid enters the metering device at the indoor coil, which converts it into low-pressure liquid. This liquid can now be used again, Figure 14-6.

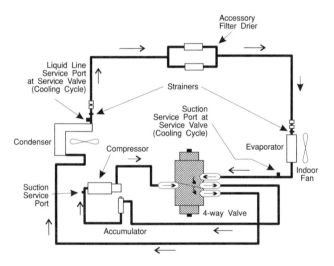

Figure 14-6. *Heat Pump Cooling Cycle*

HEATING MODE

In the heating mode, the four-way valve changes the flow direction of the gas and liquid refrigerant. In this case, the high-pressure gas moves to the indoor coil, rather than the outdoor coil. The indoor fan creates air movement which subsequently cools the high-pressure gas in the indoor coil, and the heat from this process warms the indoor air. This causes the high-pressure gas to condense into high-pressure liquid, and this liquid leaves the indoor coil and flows through the drier. (The high-pressure liquid which leaves the indoor coil bypasses the metering device at the

indoor coil.) The high-pressure liquid then enters the metering device at the outdoor coil and changes to low-pressure liquid. The liquid boils and turns into low-pressure gas. This gas then leaves the outdoor coil, enters the four-way valve, and moves through the accumulator into the compressor. Here, the compressor compresses the low-pressure gas into high-pressure gas which exits the compressor and travels to the four-way valve. The indoor coil receives this high-pressure gas and condenses it into high-pressure liquid. Figure 14-7 illustrates the heating cycle.

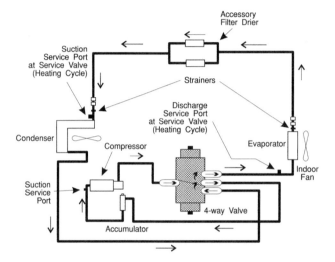

Figure 14-7. *Heat Pump Heating Cycle*

DEFROST MODE

A defrost mode is necessary, because in the heating season, it is likely the outdoor coil will operate at a temperature below freezing. This causes any moisture present in the air to freeze on the coil surface. The frost on the coil acts as an insulator, blocking heat transfer from the outdoor air to the coil. This causes the outdoor coil to lose its efficiency, and the result is the living space does not become heated.

The heat pump senses when there is a need to defrost, and subsequently, the four-way valve shifts position and starts the cooling mode. This shift causes the high-pressure gas to move to the outdoor coil, where it melts any ice or frost that is built up on the coil. Once the ice melts, the coil temperature rises, signaling the four-way valve it may reverse and return to the heating mode.

During the defrost mode, the living space does not receive any heat. For this reason, many heat pumps contain back-up electric heaters which heat the space while the defrost mode takes place. These electric heaters can be somewhat costly to operate.

TYPES OF HEAT PUMPS

As stated previously, there are three basic types of heat pumps: air-to-air, water-to-air, and ground-to-air. There is a water-to-water heat pump, but this type is not used often.

AIR-TO-AIR

The air-to-air heat pump is the most popular type of heat pump. The air-to-air heat pump is available as either a split system, or as a package unit. In a split system, the compressor and outdoor coil are placed outside, while the indoor coil and fan are housed in a cabinet (called an air handler) in the living area. The indoor coil cools or heats the return air and returns this air, now called supply air, to the living space.

The split system, air-to-air heat pump works best in a milder climate, where the winter temperature seldom drops below 0 °F. The reason being, this pump absorbs heat from the outside air. When the outside air temperature drops, the heat pump has a harder time absorbing heat.

As mentioned earlier, many heat pumps contain back-up electric heaters to heat the living space while the system is in defrost mode. In split system, air-to-air heat pumps, this heater is normally located in the air handler. These heaters also help warm the living space when the system cannot maintain the space temperature due to low outdoor air temperature.

The air-to-air pump is also available as a package unit. In this case, the indoor and outdoor coils are enclosed in a single cabinet. The duct system routes the return air to the indoor coil, and then routes the supply air to the living space. These package units are normally mounted on the roof or on the ground next to the building. These systems, too, have optional electric heat compartments.

WATER-TO-AIR

The water-to-air heat pump is a split system, with the outdoor coil resting in the water source, and the indoor coil remaining in the living space. In this type of heat pump, a lake or well may be used as the water source. Well water works particularly well in this system, because its average temperature is a constant 50 °F, plus or minus 10 °F. Lakes, streams, and ponds (surface water) do not work as well, because their temperatures fluctuate with the seasons. It is possible to use a deep lake, provided it does not freeze. In this situation, the outdoor coil should rest at the bottom of the lake, where the water remains under pressure and does not freeze.

In the water-to-air heat pump heating mode, heat is transferred from the water to the refrigerant, and then to the living space. The basic heating cycle remains the same. In the cooling mode, the heat removed from the living space air is transferred to the refrigerant and then moved back to the water source by way of a copper or plastic pipe. It is important to check local codes when returning water to its source, as some areas have strict guidelines as to where this water can go.

The one major drawback to installing a water-to-air heat pump is scale. As stated previously, scale occurs when the minerals in water deposit on a surface. Left untreated, scale reduces heat transfer in the system. To prevent or remove this scale, it is usually necessary to use chemicals. As these chemicals can pollute drinking water, it is necessary to check local codes for guidelines concerning water treatment.

GROUND-TO AIR

This type of split system heat pump is very energy efficient, however, it is rarely installed. This is often due to the installation cost, which can be very high. The ground is a good source for heat, because beneath the frost line, the ground maintains a constant temperature. This works especially well in the cooling mode, as the ground is a limitless source for heat rejection.

The ground-to-air heat pump, also called a closed-loop earth-coupled system, uses high density polyethylene pipe to connect the coils. The pipe is filled with water and/or antifreeze and then buried in the ground. The pump moves the water and/or antifreeze through the pipe. In the heating mode, the water and/or antifreeze transfers heat from the ground to the refrigerant. In the cooling mode, the water and/or antifreeze transfers heat from the system to the ground.

This system, too, has its drawbacks. For example, the underground pipes expand and contract, thus moving the soil around them. This can cause an air pocket to form around the piping which may reduce the heat transfer. Also, if a leak forms in the piping, finding and repairing the leak can be costly.

AIR FLOW

In an air conditioning system, the standard air flow should be 400 cfm/ton over the evaporator and 700 cfm/ton over the condenser. In heat pumps, however, the coils must function as both condenser and evaporator, so the air flow over each must be equal. This standard air flow for a heat pump is 450 cfm/ton. This is a greater volume of air over the inside coil than on the equivalent in a straight air conditioning system. Therefore, if a heat pump is used to replace an existing system, the duct system may be undersized. Also, much less air flows over the outside coil compared to standard air conditioning. This means the heat rejection during the cooling mode is less efficient, while during heating, the heat pickup is better. Some systems use a two-speed motor on the outside coil. This delivers more air in the cooling mode which then holds down the system pressure and allows the unit to operate more efficiently in the cooling mode when more air is needed over the outside coil.

RATINGS

The selection of a heat pump for a given structure is made on the basis of its cooling and heating capacities, which can be calculated using the coefficient of performance and energy efficiency ratio methods. The heat pump should not be oversized by more than 1/2 ton, or 6000 Btuh, even though additional capacity may be needed in the heating mode. As with air conditioning, control of latent heat is lost with an oversized unit because of shorter running time, and undersizing does not handle the load.

Additional heating requirements are supplied by supplemental electric resistance heaters, in 5 to 10 kW increments. The number of electric heaters required is determined by the calculated heat loss of the structure. These electric resistance heaters are often called strip heaters, Figure 14-8.

Figure 14-8. Electric Resistance Heater. Courtesy, Carrier Corporation, a Subsidiary of United Technologies Corporation.

COEFFICIENT OF PERFORMANCE

The coefficient of performance (COP) is the method used to measure the energy consumed by the heat pump as opposed to the amount of energy produced by the heat pump. The COP is the standard used to compare different brands of heat pumps.

To figure out the COP, it is necessary to understand electric heat. When a building uses electric heat, the building receives 1 Watt of usable heat for every 1 Watt of energy sent by the power company (1 Watt of electricity equals 3.413 Btu). Therefore, electric heat is 100 percent efficient, meaning the output is the same as the input, or the COP is a ratio of 1 to 1. When an air-to-air heat pump is installed, this efficiency may improve as much as 3 to 1. The COP is calculated by multiplying the amount of energy consumed (in Watts) by 3.413; then, this number is divided into the Btuh output.

To illustrate, a unit operating at 20 °F with a 26,000 Btu capacity uses 2,700 Watts to accomplish its output. To calculate the COP, it is necessary to multiply 2,700 by 3.413. This number is the input. Then, the output, 26,000 is divided by the input (9,215). The answer, 2.82, is the COP. This means the heat pump, in the heating mode, produces 2.82 times more heat output (Btuh) than its electrical input in Btuh.

There is always a certain amount of heat in the outside air, even at 0 °F. However, as the temperature decreases, it becomes more difficult for the heat pump to extract the heat from the air. This causes the COP to fall proportionately. While the COP may fall, it is normally still more efficient in its consumption of electric power than its electric heat counterpart. Even at an outside temperature of 0 °F, the typical air-to-air heat pump has a COP of 1.5. When the temperature becomes extremely cold, the COP can drop to 1.0, meaning that for every Watt used, an equal amount of Btu is produced. The normal range for a heat is usually 2 to 3 COP.

BALANCE POINT

The balance point is the lowest outside temperature at which a heat pump can operate and still adequately heat the living space. At the balance point, the heat pump must run continuously in order to keep the living space heated. When the temperature drops below this point, the heat pump runs continuously, but it is not able to adequately heat the living space. Should this occur, some form of back-up heat is required, because in cold weather, a structure leaks heat quickly. Figure 14-9 shows the structure heat loss versus the heating unit capacity. The point at which the two lines intersect is the balance point.

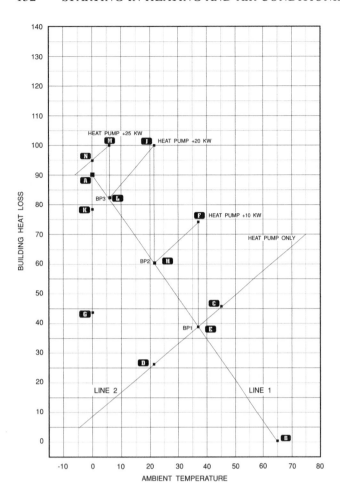

Figure 14-9. *Balance Point Graph*

ENERGY EFFICIENCY RATIO

The COP measures only heating, and heat pumps provide cooling and heating. The energy efficiency ratio (EER) is used to measure the efficiency of air conditioners, heat pumps, etc., in the cooling mode. EER is the ratio of the energy produced directly in Btu to the energy consumed in Watts; the higher the number, the more efficient the product. In the case of EER, the number calculated should be much higher than the COP. In heat pumps, an EER of eight or nine is considered good.

HEAT PUMP COMPONENTS

Some of the components discussed in this section are found on all heat pumps, while others occur on only a few heat pumps. This section is meant to provide a brief overview of the components most likely encountered when working with a heat pump.

A low voltage (24 V) circuit controls heat pump operation. Power for this circuit originates from the transformer secondary, where line power is reduced to 24 V. This low voltage is wired to the indoor thermostat, which functions as the signal center for all component operation.

ACCUMULATORS

Accumulators are necessary to protect the compressor from low outdoor temperatures. When a heat pump unit shifts from one mode to the other, especially in the defrost mode, a large surge of liquid refrigerant returns to the compressor. In order to minimize slugging, a suction line accumulator must be installed to receive this liquid before it reaches the compressor. The accumulator stores this refrigerant until the temperature increases. Then, the liquid boils off and returns to the system in gas form. The accumulator also acts as a storage point for refrigerant during the heating cycle, when the system requires much less refrigerant.

The accumulator is located in the suction line between the compressor and the four-way valve. Normally, an accumulator can hold 50 percent of the charge in a system.

CHECK VALVES

Check valves route the refrigerant flow through the metering device and bypass the metering device not in use. Check valves also allow refrigerant to flow in only one direction. An arrow located on the check valve body indicates the flow direction of the valve.

When a thermostatic expansion valve (TXV) is the metering device used, it is often coupled with a check valve. A combination of one TXV and one check valve is normally located at each coil, indoor and outdoor. In the cooling mode, one check valve forces liquid refrigerant flow to the indoor TXV, while the other check valve bypasses liquid refrigerant around the outdoor TXV. Conversely, in the heating mode, one check valve forces liquid refrigerant flow to the outdoor TXV, while the other check valve bypasses liquid refrigerant around the indoor TXV.

When a capillary tube is used, it is also often coupled with a check valve. In the heating mode, the check valve forces refrigerant through a second capillary tube. This second capillary tube restricts the amount of refrigerant entering the outdoor coil. The restriction is necessary, because most heat pumps use less refrigerant in the heating mode as opposed to the cooling mode.

COMPRESSORS

The same principles apply to both heat pump compressors and normal air conditioning compressors, with a few exceptions. Problems that occur with a normal air conditioning compressor also normally apply to a heat pump compressor; however, the manner in which the service technician remedies these problems may differ. As always, it is important to consult the manufacturer's literature concerning the heat pump compressor before any service is performed.

A heat pump compressor works harder than the compressor found in a normal air conditioning system, because it is expected to perform in both heating and cooling modes. Also, a heat pump compressor in the heating mode must produce a compression ratio almost double to what it produces in the cooling mode. This increase in compression is necessary, because when in the heating mode, the operating pressure is lower, which can cause liquid refrigerant to enter the compressor. For this reason, the heat pump compressor must move a very large volume of gas when in the heating mode. The suction pressures are also lower when in the heating mode, so the unit runs hotter, as less suction gas is available for cooling. For these reasons, it is very important to only replace a heat pump compressor with another heat pump compressor recommended by the manufacturer. Compressors from normal air conditioning systems should not replace heat pump compressors.

Most heat pump compressors in residential use are reciprocating compressors (reciprocating compressors were discussed in Chapter 10). The only difference between reciprocating compressors in heat pumps versus normal air conditioning systems is that some heat pump reciprocating compressors have heavier valves. Also, these compressors may have thicker connecting rods and greater bearing surfaces. These slight changes are necessary to compensate for various dangers a heat pump compressor might incur during normal operation.

CONNECTING LINES

For the most part, the piping used in heat pump systems is the same as is used in normal air conditioning systems. That is, the piping comes in line sets which can either be nitrogen-charged or precharged. The same precautions should be followed when installing heat pump systems as are followed when installing a normal air-conditioning system.

One difference between heat pump systems and normal air conditioning systems is that heat pump systems contain a large-diameter vapor line. During the winter months, the vapor line may be 200 °F, so thicker insulation in this line is usually necessary. The vapor line should not be located near other objects which may be affected by the vapor line's temperature.

CRANKCASE HEATERS

The crankcase heater provides heat to the compressor crankcase at all times, regardless of whether or not the unit is running. The crankcase heater is the most important safety device in the heat pump, because it keeps the refrigerant from mixing with the oil when the unit is not running. This refrigerant migration was discussed earlier. To review, when refrigerant and oil mix, foaming occurs, which reduces oil lubrication. Also, slugging can occur, because the refrigerant reaches the compressor once it mixes with the oil. The crankcase heater guards against refrigerant migration and slugging.

The compressor is normally located outdoors in split systems, so in this configuration, the compressor operates in a cold environment. The cold environment can speed refrigerant migration, thus increasing the risk of compressor failure. It is important for the service technician to warn the heat pump owner to leave the crankcase heater on at all times to ensure a properly-functioning compressor.

The compressor shell houses the crankcase heater, which is nothing more than a high resistance wire. There are also strap-on crankcase heaters that strap around the compressor shell. In order to make sure the crankcase heater is operating properly, it is necessary to check its ohms. If the ohm reading is infinity, the heater is burned out and needs to be replaced. A normal crankcase heater shows high resistance on the meter.

DEFROST CONTROLS

The defrost controls, as the name implies, control the defrost cycle. These controls are made up of the defrost thermostat, time clock, and relay. Sometimes, a high-pressure control takes the place of the defrost thermostat.

Defrost Thermostat. The defrost thermostat usually either starts or ends the defrost cycle. Or, this thermostat can perform both starting and ending functions.

Normally, the defrost thermostat is located on the outdoor coil or liquid line. When the outside temperature falls below a certain preset point, the thermostat closes. This causes the time clock to call for defrost. When the outdoor coil warms up enough to melt the ice, the thermostat opens and the defrost cycle ends.

The defrost thermostat is normally wired into the defrost time clock and wired in series with the defrost relay.

Defrost Time Clock. The time clock, itself, does not start the defrost cycle, that is the job of the defrost thermostat. The time clock starts the defrost sequence once the defrost thermostat closes.

The time clock contains contacts, which, after a preset period of time, close for a short while in order to start the defrost cycle. The contacts can only close if the defrost thermostat closes first. The time clock normally checks for defrost every 30, 60, or 90 minutes, less if more frequent defrost is required.

If the system does go into defrost, the cycle terminates when the coil temperature reaches 65 °F. When ice accumulation is present at an outdoor temperature of 40 °F, the defrost cycle may take 5 to 6 minutes. At lower temperatures (and thus lower moisture content) such as 10 °F, the defrost cycle may only take 3 minutes. In most systems, if high wind or low outdoor temperature is present, the time clock can override the thermostat and terminate defrost after 10 minutes, even if the coil does not reach 65 °F.

Defrost High Pressure Control. Sometimes a defrost high pressure control, called an air pressure switch, takes the place of the thermostat-time clock system. When this is the case, the air pressure switch is not wired into a time clock.

The air pressure switch checks pressure, as opposed to the thermostat which checks temperature. The outdoor coil pressure is lower than normal when it is iced over. The air pressure switch senses this pressure change and closes, energizing the defrost relay and starting the defrost cycle.

In this type of system, high winds can cause the air pressure switch to not function properly. Also, a dirty outdoor coil can start the defrost cycle. For this reason, the air pressure switch is usually coupled with a defrost thermostat. The defrost thermostat acts to back up the air pressure switch. For example, when the air pressure switch senses a defrost cycle is needed, the defrost thermostat takes a temperature reading and determines whether or not the temperature coincides with the need for a defrost cycle.

Typical pressure readings, in inches W.C., are 0.2 inches for a clean coil, 0.5 inches for a coil with 50 to 75 percent ice coverage, and 0.65 inches for a coil completely covered with ice. The control can be set, using an adjusting screw, to initiate defrost anywhere within a range of 0.15 to 1.0 inches W.C. Since there is no timer involved with this system, the sensor initiates defrost only when it is required.

Defrost Relay. The defrost relay is wired to both the defrost thermostat-time clock system and the high-pressure control system, depending on which system is in place. In order for the defrost relay to be energized, a complete circuit must be made. A complete circuit occurs when either the defrost thermostat or air pressure switch closes.

The defrost relay contains a magnetic coil, as well as normally-closed and normally-open contacts. The coil is usually low voltage and the contacts are line voltage. When the relay is energized, its normally-closed switches open, causing the outdoor fan to turn off and the four-way valve to shift. Also, the normally-open contacts in the relay close, causing the back-up electric heater to energize during defrost.

The defrost relay coil can either be 21 vdc or 24 vac.

FAIR WEATHER SWITCH

The fair weather switch protects the heat pump in the cooling mode. It is not a good idea to run the heat pump if the outdoor temperature is low, as this can cause liquid refrigerant to flood the compressor. The fair weather switch guards against this occurring.

The fair weather switch senses high pressure and is connected to the discharge line. The switch is wired in series with the outdoor fan motor. When outdoor temperature falls to a point too low to maintain proper pressure, the fair weather switch opens and shuts off the outdoor fan. Without the fan operating, no air can flow over the outdoor coil. When this happens, the high-pressure side pressure rises, the fair weather switch closes, and fan operation is restored.

Fair weather switches normally open at 175 psig and close at 250 psig.

INDOOR THERMOSTAT

The indoor thermostat is described later in this section.

LIQUID LINE FILTER DRIER

This filter driver is located in the liquid line and absorbs any moisture that is present in the system. The liquid line filter drier on a heat pump is different from driers used in normal air conditioning, because heat pump driers must be able to handle refrigerant flow in two directions. The normal air conditioning drier only handles refrigerant flow in one direction. Liquid line filter driers improve the performance ability of heat pumps.

LOW AMBIENT LOCKOUT

The low ambient lockout is an outdoor thermostat that shuts off the compressor when the outside temperature drops below a preset temperature. Depending on the unit, this present temperature can be as high as 20 °F or as low as 0 °F or below.

As stated previously, when the outdoor temperature becomes too low, the heat pump loses efficiency. Should the temperature continue to drop, it is possible the heat pump may not be able to provide enough heat for the living space. For this reason, the low ambient lockout shuts off the compressor. If the compressor continues to run in very cold weather, it may wear itself out.

The low ambient lockout is not available on every compressor. When a low ambient lockout is not included on the compressor it is usually because the manufacturer recommends running the compressor at all times. It is important to check the manufacturer's literature to ensure this is the case.

METERING DEVICES

In a heat pump, a variety of metering devices may be used, including expansion valves and capillary tubes. In some heat pumps, a combination of two different metering devices may be used. The main function of the metering device, however, is the same as it is in a normal air conditioning system; that is, to regulate refrigerant flow, so the correct amount is always present in the system. A metering device also works to change high-pressure liquid into low-pressure liquid.

Thermostatic Expansion Valves (TXV). As stated previously, the TXV is often used in conjunction with a check valve. In this system, a check valve and TXV are placed at each coil, thus forcing the liquid refrigerant into the valve, so it can change the high-pressure liquid into low-pressure liquid. The low-pressure liquid can then be used for evaporation.

The TXV responds to changing refrigerant temperature and pressure by way of a remote refrigerant-charged bulb and an equalizer tube. The remote bulb senses the temperature of the gas leaving the evaporator and the equalizer tube senses the pressure of the refrigerant leaving the evaporator.

Another type of TXV is known as a bi-flow TXV. When the bi-flow valve is used, it operates as a regular TXV in the cooling mode. In the heating mode, however, the refrigerant flow is reversed. The bi-flow TXV has a check valve located internally, which allows the reversed refrigerant flow to pass through it.

Capillary Tube. A capillary tube, as in a normal air conditioning system, is a constant feed device and does not respond to differences in the system. The capillary tube does allow refrigerant flow in either direction, however, it is not used often, because different sizes of capillary tubes are often required for cooling and heating modes. One method of using a capillary tube is to combine it with a check valve to reverse the flow. When this method is employed, two capillary tubes are normally used.

MILD WEATHER SWITCH

The mild weather switch protects the heat pump in the heating mode. If the heat pump is in the heating mode when the outside air temperature is over 70 °F, very high discharge pressure results. The mild weather switch protects the heat pump against this high pressure.

The mild weather switch senses discharge pressure in the heating mode and is located in the discharge line, between the indoor coil and the four-way valve. When warm outdoor temperature occurs, the mild weather switch opens, turning off the outdoor fan. This causes the air flow over the outdoor coil to decrease and the high-pressure side pressure also decreases. This decrease prompts the switch to close, which turns on the outdoor fan. Normal fan operation then resumes.

OUTDOOR THERMOSTAT

The outdoor thermostat is mounted outdoors and has a remote bulb or coil which senses the outdoor air temperature. This thermostat is used to energize the back-up electric heater when the outdoor temperature falls dramatically. This thermostat then regulates the back-up electric heater, bringing in additional stages of electric heat before the living space becomes uncomfortable. When the outdoor temperature rises, the outdoor thermostat returns control of the heating operation to the indoor thermostat.

SUCTION LINE FILTER DRIER

This filter drier is normally installed in the suction line, which is the center line coming out of the four-way valve. This line then runs to the compressor. The suction line filter drier removes acid and contaminants from the system when a burnout occurs. This filter drier also traps moisture.

HEAT PUMP INSTALLATION

The manufacturer of the heat pump normally includes installation instructions concerning piping, wiring and air flow clearances. While these instructions are usually similar to a normal air conditioning installation, it is still necessary to follow the heat pump instructions.

| Terminal Identification for a Typical Heat Pump Thermostat ||
Terminal Marking:	Function:
R or V	Power from the low voltage side of transformer
Y_1 or C_1	Power when COOLING contacts close to low voltage contactor coil which closes high voltage contacts to bring on compressor circuit and allied components.
Y_2 or C_2	Second Stage Cooling
W_1 or H_1	Power when HEATING contacts close to reversing valve relay which can energize or de-energize reversing valve solenoid and energize compressor circuit. {First Stage Heating}
W_2 or H_2	Second set of contacts closing energize auxiliary heat strips for additional heat or during defrost. {Second Stage Heating}
G or F	Power to indoor fan coil to close high voltage, contacts to turn on indoor fan.
O or D	*There is power available at the O terminal at all times when the manual switch at the thermostat is set for cooling.* If reversing valve relay is wired to O, then reversing valve solenoid will be energized in cooling at all times.
B or Z	*There is power available at the B terminal at all times when the manual switch at the thermostat is set for heating.* If reversing valve relay is wired to B, then reversing valve solenoid will be energized in heating at all times.
X or L	Warning light monitor.
E	Emergency heat control.

Figure 14-10. Terminal Identification. Courtesy, Richard Jazwin, Heat Pump Systems & Service, *Published by Business News Publishing Company.*

Heat pumps contain drain holes or slots in the cabinet which drain water formed in the defrost cycle. Defrost occurs in the winter months, so these drain holes must be free of obstruction. If an obstruction is present, ice can form inside the unit. These drain holes should be checked every month, as well as after heavy snow or ice storms.

When installing a heat pump outdoors, the normal practice is to use a pad or mounting frame to raise the unit at least 6 inches off the ground. In areas where heavy snowfall is likely, it is necessary to increase the mounting height to 20 or 24 inches. This is accomplished by mounting the unit on cement blocks. Gravel placed around the unit is also suggested, in order to aid in water drainage.

It is necessary to install the unit in a reasonably open area where drifting snow cannot pile up around it. Also, the prevailing winds should not blow directly through the coil section, which can be difficult when the unit is round or three-sided. As with a normal air conditioner condenser, there should be enough clearance from the building so water or ice from the roof cannot drop directly on or in front of the coil.

HEAT PUMP THERMOSTATS

A heat pump thermostat differs from thermostats in other systems. This is because in a heat pump thermostat, there are two complete heating systems and one cooling system. Two heating systems are necessary, because the auxiliary heat (i.e., back-up electric heaters) becomes the primary heat source when the heat pump fails. For this reason, the thermostat identifies the auxiliary heat as its own system.

Figure 14-10 is provided to help understand the schematic wiring diagrams in this section. This figure shows the terminal identification for a typical heat pump thermostat. In some heat pumps, the terminal identification is different, so it is necessary to check the manufacturer's literature first.

Heat pump thermostats are normally of the two-stage heating and two-stage cooling variety, Figure 14-11. There are variations in the thermostat, but these variations usually concern the number of stages of heating or cooling.

The two heating stages are known as the first stage and second stage. In the first stage, the heating contacts close,

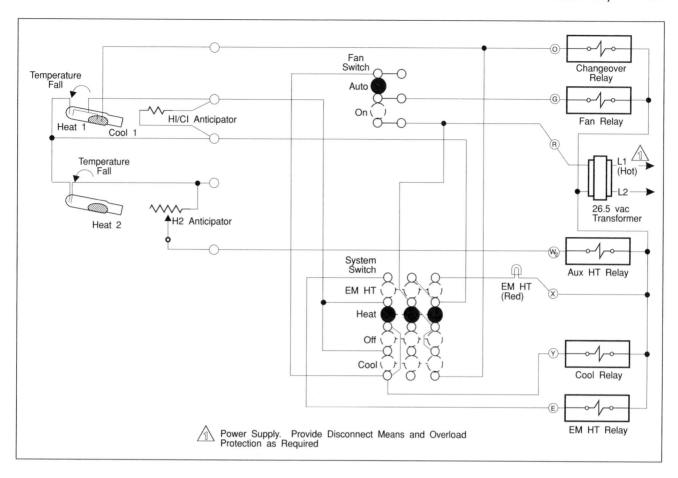

Figure 14-11. Heat Pump Thermostat with Two-Stage Heating and Two-Stage Cooling

and the system supplies heat to the living space. The second stage turns on the back-up electric heaters when more heat is required. The cooling mode also has a first and second stage. The first stage uses the heat pump itself to cool the living area. The second stage operates the compressor.

Heat pump thermostats have fixed cooling anticipators. They do have adjustable heat anticipators, which apply to the first or second stage of heating, or a combination of the two. The value is about 0.4 A for the first stage and 0.5 A for the second stage.

As in a normal air conditioning thermostat, a subbase is needed in the heat pump thermostat for the various switching functions. A heat pump thermostat often includes an emergency heat switch which energizes the electric heaters in case of compressor failure, or while the compressor is shut down for service. This is a manual switch that bypasses the other controls. An indicator light reminds the owner when electricity alone is heating the living space.

A thermostat can either have a manual or automatic changeover from heating to cooling. In a manual thermostat, a switch can be moved to the heat or cool position.

Once a mode is selected, either opening-closing contacts, or a mercury bulb controls unit operation. Automatic thermostats contain a temperature-sensing element, such as a bimetal. This element then controls mercury bulb contacts. In both systems, the indoor fan may be operated continuously, or set to automatic, which cycles the fan on and off with the compressor.

THERMOSTAT IN COOLING MODE

In the examples that follows, it is assumed the thermostat contains an automatic changeover and the fan is set to automatic.

In the cooling mode, the first-stage contacts close, energizing the four-way valve, Figure 14-12. Once the temperature in the living area rises 1 °F, the second-stage contacts close, energizing the compressor contactor and the indoor fan relay. This starts the compressor, outdoor fan, and indoor fan, Figure 14-13.

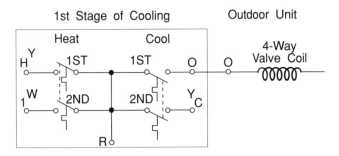

Figure 14-12. *Cooling Mode, First Stage*

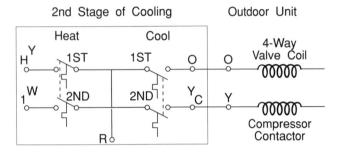

Figure 14-13. *Cooling Mode, Second Stage*

Once the compressor starts removing heat from the living area, the temperature in the living area falls. When the temperature reaches the desired point, the second-stage contacts open, and the compressor, outdoor fan and indoor fan stop. However, the first-stage contacts remain closed, and the four-way valve remains energized. When the temperature in the living area begins to rise again, the second-stage contacts close once again, starting the compressor and the outdoor and indoor fans.

When the outdoor temperature becomes cooler, the second-stage contacts remain open, as the temperature in the living area remains cool. This causes the first-stage contacts to open, thus de-energizing the four-way valve. When the unit starts up again, it will be in the heating mode.

THERMOSTAT IN HEATING MODE

When the temperature of the living space falls, the first-stage heating contacts close, starting the compressor,

outdoor fan, and indoor fan, Figure 14-14. This action does not energize the four-way valve, so the hot gas from the compressor moves toward the indoor coil. Once the temperature in the living space reaches a desired level, the first-stage heating contacts open, and the compressor, outdoor fan and indoor fan stop.

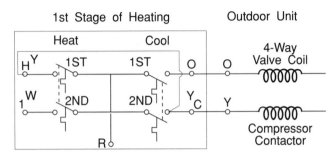

Figure 14-14. *Heating Mode, First Stage*

If the temperature of the living area continues to drop due to falling outdoor temperatures, the heat pump cannot continue to heat the living area on its own. When this occurs, the second-stage heating contacts close, starting the back-up electric heaters, Figure 14-15. Once the temperature in the living space reaches the desired level, the second-stage heating contacts open, turning off the back-up electric heaters.

As stated previously, outdoor thermostats are often wired in series with the back-up electric heaters. This prevents all the auxiliary heat from coming on with the second stage. Instead, the heat delivered to the living space is based on the outdoor thermostat temperature reading.

THERMOSTAT IN EMERGENCY HEAT MODE

On most thermostats, there is an emergency heat switch which can be operated manually. Emergency heat is only to be used when the heat pump cannot function. When turning the switch to emergency heat, the compressor stops, energizing the back-up electric heater. In this case, the indoor thermostat, not outdoor thermostat, controls the electric heater.

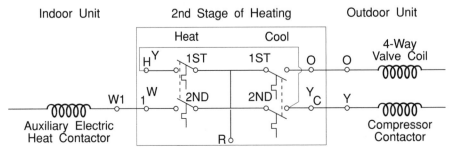

Figure 14-15. *Heating Mode, Second Stage*

THERMOSTAT IN DEFROST MODE

The defrost mode is basically the same as the cooling mode. The defrost controls were described previously, and these controls determine whether or not the defrost mode is needed.

HEAT PUMP SERVICE

Because heat pumps operate all year long through both winter and summer, heat pumps require more service than a singular heating or air conditioning system. Also, as there are more components located on a heat pump that are not present on a singular heating or air conditioning system, there is an increased possibility of a system malfunction.

As always, it is necessary to thoroughly read the manufacturer's literature before any service is performed. The literature offered from manufacturers usually includes installation instructions, electrical schematics, and service sections. To successfully service a heat pump system, the service technician must also take the time to learn the sequence of operation for the particular heat pump which requires service.

The service procedures that follow may not apply to every heat pump brand; however, some problems are universal and do apply to every brand. It is also important to keep in mind that while some heat pump brands may be better than others, all can be subject to failure. This section assumes the system has performed satisfactorily in the past, and now has a service problem. The various heat pump components were discussed previously, and their functions should be reviewed before attempting service on a unit. The charts shown at the end of the section can be a guide to service procedures, but it is necessary to think the problem out before proceeding with a correction.

GENERAL SERVICE

Before servicing a component, it is first necessary to check the air pump generally, to ensure the problem is not restricted air or water flow. Also, the filters must be clean, the fan motors oiled, the duct system adequate, and the temperature rise (or fall) across the coil correct for the unit to operate properly.

It is then necessary to ask several question which will help to locate the problem. These questions include: in which mode does the problem occur, heating or cooling? is the unit defrosting properly, and if not, is this causing the problem? what type of defrost system is in place on the heat pump? is there really a malfunction, or is the owner simply not used to the way in which a heat pump works?

Before starting the service check, check the items in the following list:

- The discharge line temperature should not exceed 225 °F.
- In the heating mode, the suction pressure should be the wet-bulb temperature minus 26 °F. In the cooling mode, the suction pressure should be the wet-bulb temperature minus 24 °F.
- The suction line temperature should not exceed 70 °F.
- In the heating mode, the air temperature at the coil outlet should be between 85 °F and 105 °F. When the outdoor air is cooler, the outlet temperature should be closer to 85 °F, and when the outdoor air is warmer, the temperature should approach 105 °F.
- In the cooling mode, the difference in the coil air temperature between the supply air and the return air should be between 18 °F and 22 °F.

CHECKING ELECTRICITY

Most heat pump trouble shows up initially as an electrical problem, so the analysis starts at this point.

Indoor and Outdoor Disconnect Switches

1. Check that supply voltage to each disconnect is 240 V.
2. Check for blown fuses. Replace any blown fuses. Before applying power, locate and repair the short in the system which initially blew the fuse.

Transformers

1. Check that input voltage is 240 V.
2. Check that secondary voltage is 24 V.
3. If either voltage is not correct, disconnect all wires, turn off power and check for shorts (infinite continuity) and resistance coil.

Relays

1. Check coil for voltage input.
   CAUTION: Some relays are 240 V.
2. Check across each contact for voltage. A voltmeter reading means the contact is open. No voltage means the contact is closed.
3. Disconnect all wires, turn off power and check for shorts. Check coil for resistance.

NOTE: Some wiring schemes do not have a connection to one of the DPDT terminals, so these cannot be checked by voltage. Look at the wiring diagram, and check for continuity.

Outdoor Thermostat

1. Turn power off.
2. Check continuity across contacts (should be open if outdoor air is above dial setting).
3. Turn dial below outdoor air setting. Check for continuity across contacts; these should be open. If they are not, replace them.

Indoor Thermostat

1. Check that thermostat is level.
2. With power off and thermostat not calling for heating or cooling, check each circuit for continuity—all should be open.
3. With power on, check the heating and cooling circuits for voltage. No voltage indicates circuit is closed, and a reading of 24 V means the circuit is open.
4. Check heat anticipators in both first- and second-stage heating, and set to the readings recorded. Note that some first-stage anticipators and all cooling anticipators are fixed.

Four-Way Valve

1. Check the compressor discharge pressure. Low pressure will not operate the four-way valve.
2. Check the compressor suction pressure. High suction pressure indicates a leaking check valve.
3. Perform a leak test, and recharge the unit if needed. Normal suction pressure and low discharge pressure may indicate a defective compressor.
4. Energize the four-way valve. Remove the locknut to free the solenoid coil in the valve. Slide the coil partly off the stem to see if there is a magnetic force trying to hold it on.

CAUTION: Some coils are 240 V.

5. Listen for a clicking noise as the coil moves. This indicates the plunger is in the non-energized position. Replace the coil on the stem and listen for another click. This indicates the valve is changing over.
6. Check for physical damage to the valve. Inspect for dents, deep scratches or cracks.

When the above checks have been made satisfactorily, then perform the Touch Test, Table 14-1.

CHECKING THE DEFROST MODE

Before checking the components needed for defrost mode, it is necessary to first check the heating and cooling modes. If the heating and cooling modes are operating satisfactorily, and a problem still exists, it can usually be attributed to the defrost mode components.

In general, when a system is not defrosting, or if the system is defrosting erratically, the defrost timer is normally at fault. When the system does not perform in the defrost mode, this usually means the timer motor is faulty, or the contacts in the timer are stuck open. When the defrost mode operates erratically, this normally indicates the contacts are stuck closed.

To check the defrost components, it is necessary to consult the manufacturer's literature concerning defrost on a particular heat pump. This is because there are various defrost components used, depending on the heat pump. The service procedures change, depending upon which components are used.

CHECKING CHECK VALVES

The usual check valve has a steel ball bearing that opens under pressure from one direction and closes under pressure from the opposite direction. Its operation can be checked in the following ways:

1. With the unit off, cycle the ball bearing with a large magnet. Listen for the sound of the ball closing.
2. Check the temperature of the check valve by touching it when the unit is operating. If the valve is closed, there is a definite temperature difference across the valve. No temperature difference indicates the valve is wide open. If the valve is leaking, the temperature difference will be much less than normal. Use good measuring devices when checking for this problem.

CHARGING HEAT PUMPS

Usually, this is the most confusing area in heat pump service. This is because the heat pump requires a different quantity of refrigerant in the heating and cooling modes. In the cooling mode, the heat pump requires more refrigerant than when the unit is in the heating mode.

Proper refrigerant charge in a heat pump is a major factor in its efficient operation. Many units are installed with precharged line sets, so the proper charge is already available. These line sets are connected in essentially the same way as air conditioning line sets. Consequently, there should be no question of proper charge when connecting the heat pump for the first time, whether summer or winter. However, on a service call, the situation may be more difficult. The unit may have lost charge, or it may be necessary to remove the charge before servicing.

The most accurate and efficient way to recharge the system is to weigh in the refrigerant, following the manufacturer's instructions and specifications. This method ensures a correct charge, regardless of the ambient temperature. In all new heat pump units, manufacturers include charging charts. These charts specify the pressures needed to operate the unit in heating and cooling modes. These charts are compiled based on suction line temperature, outdoor temperature, suction pressure, superheat, subcooling, indoor wet- and dry-bulb temperatures, metering devices, and air flow over the indoor coil, Figures 14-16 and 14-17.

Each manufacturer uses different values, and these values are calculated differently by each manufacturer. For this reason, it is extremely important that only the manufacturer charts included with the heat pump unit be followed. If these charts are not available, or if the unit is older and did not come with these charts, the manufacturer's literature must be consulted, and the charge must be weighed in.

Before a charge can be weighed in, the system must be evacuated. When evacuating a system, the same basic rules apply as were discussed earlier. The charge can then be weighed by methods also discussed in a previous chapter.

No single rule exists for how to charge a heat pump. Instead, it is always necessary to obtain the manufacturer's literature concerning the charge of a particular heat pump. Guessing should never be used when charging a system, as this can lead to improper charging.

CHECKING OTHER COMPONENTS

Components common to normal air conditioning systems, i.e., compressors, capacitors, blower motors, contactors, etc., plus the electric heat elements with their limits and fusible links, are checked in the manner described in foregoing chapters.

OVERALL TROUBLESHOOTING

The following charts provide a logical troubleshooting sequence for both the heating and cooling modes. It is not necessary to memorize these charts, as most manufacturers issue similar service materials. If the manufacturer's troubleshooting chart is available, the service technician should always follow it.

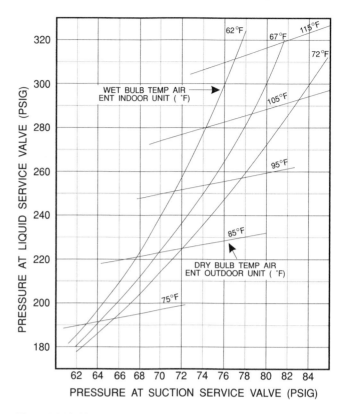

Figure 14-16. *Charging Chart 1*

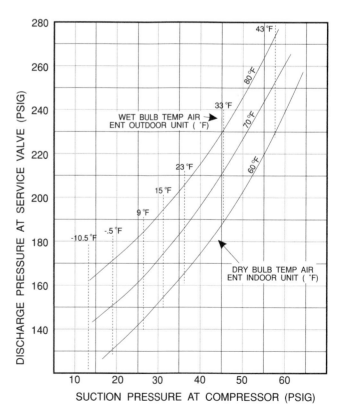

Figure 14-17. *Charging Chart 2*

VALVE OPERATING CONDITION	DISCHARGE TUBE from Compressor	SUCTION TUBE to Compressor	Tube to INSIDE COIL	Tube to OUTSIDE COIL	LEFT Pilot Back Capillary Tube	RIGHT Pilot Front Capillary Tube	NOTES: * Temperature of Valve Body. ** Warmer than Valve Body.	
	1	2	3	4	5	6	Possible causes.	Corrections
NORMAL OPERATION OF VALVE								
Normal COOLING	Hot	Cool	Cool as (2)	Hot as (1)	*TVB	*TVB		
Normal HEATING	Hot	Cool	Hot as (1)	Cool as (2)	*TVB	*TVB		
MALFUNCTION OF VALVE								
Valve will not shift from cool to heat	Check electrical circuit and coil						No voltage to coil.	Repair electrical circuit
							Defective coil.	Replace coil
	Check refrigeration charge						Low charge.	Repair leak, recharge system.
							Pressure differential too high.	Recheck system.
	Hot	Cool	Cool as (2)	Hot as (1)	*TVB	Hot	Pilot valve okay. Dirt in one bleeder hole.	Deenergize solenoid, raise head pressure, reenergize solenoid to break dirt loose. If unsuccessful, remove valve, wash out. Check on air before installing. If no movement, replace valve, add strainer to discharge tube, mount valve horizontally.
							Piston cup leak.	Stop unit. After pressures equalize, restart with solenoid energized. If valve shifts, reattempt with compressor running. If still no shift, replace valve.
	Hot	Cool	Cool as (2)	Hot as (1)	*TVB	*TVB	Clogged pilot tubes.	Raise head pressure, operate solenoid to free. If still no shift, replace valve.
	Hot	Cool	Cool as (2)	Hot as (1)	Hot	Hot	Both ports of pilot open. (Back seat port did not close.)	Raise head pressure, operate solenoid to free partially clogged port. If still no shift, replace valve.
	Warm	Cool	Cool as (2)	Warm as (1)	*TVB	Warm	**Defective compressor.**	
Start to shift but does not complete reversal	Hot	Warm	Warm	Hot	*TVB	Hot	Not enough pressure differential at start of stroke or not enough flow to maintain pressure differential.	Check unit for correct operating pressures and charge. Raise head pressure. If no shift, use valve with smaller ports.
							Body damage.	Replace Valve.
	Hot	Warm	Warm	Hot	Hot	Hot	Both ports of pilot open.	Raise head pressure, operate solenoid. If no shift, replace valve.
	Hot	Hot	Hot	Hot	*TVB	Hot	Body damage	Replace valve.
							Valve hung up at mid-stroke. Pumping volume of compressor not sufficient to maintain reversal.	Raise head pressure, operate solenoid. If no shift, use valve with smaller ports.
	Hot	Hot	Hot	Hot	Hot	Hot	Both ports of pilot open.	Raise head pressure, operate solenoid. If no shift, replace valve.
Apparent leak in heating	Hot	Cool	Hot as (1)	Cool as (2)	*TVB	**WVB	Piston needle on end of slide leaking.	Operate valve several times then recheck. If excessive leak, replace valve.
	Hot	Cool	Hot as (1)	Cool as (2)	**WVB	**WVB	Pilot needle and piston needle leaking.	Operate valve several times then recheck. If excessive leak, replace valve.
Will not shift from heat to cool	Hot	Cool	Hot as (1)	Cool as (2)	*TVB	*TVB	Pressure differential too high.	Stop unit. Will reverse during equalization period. Recheck system.
							Clogged pilot tube.	Raise head pressure, operate solenoid to free dirt. If still no shift, replace valve.
	Hot	Cool	Hot as (1)	Cool as (2)	Hot	*TVB	Dirt in bleeder hole.	Raise head pressure, operate solenoid. Remove valve and wash out. Check on air before reinstalling, if no movement, replace valve. Add strainer to discharge tube. Mount valve horizontally.
	Hot	Cool	Hot as (1)	Cool as (2)	Hot	*TVB	Piston cup leak.	Stop unit, after pressures equalize, restart with solenoid deenergized. If valve shifts, reattempt with compressor running. If it still will not reverse while running, replace valve.
	Hot	Cool	Hot as (1)	Cool as (2)	Hot	Hot	Defective pilot.	Replace valve.
	Warm	Cool	Warm as (1)	Cool as (2)	Warm	*TVB	**Defective Compressor.**	

Table 14-1. *Courtesy, Ranco North America.*

REVIEW QUESTIONS

1. What is the difference between an air-to-air heat pump, a water-to-air heat pump, and a ground-to-air heat pump?
2. How are heat pumps rated?
3. Explain what COP means.
4. Explain the meaning and importance of balance point.
5. What is the purpose of an accumulator?
6. What does the four-way valve do?
7. What do the heat pump coils do?
8. What happens to the compressor when the four-way valve changes over?
9. What makes a four-way valve change position?
10. Why does a heat pump need a defrost cycle?
11. How does a defrost timer work?
12. What is an air pressure switch?
13. What special installation procedures are necessary for a heat pump?
14. What is the purpose of the defrost relay?
15. What happens when the system goes into the defrost mode?
16. What happens when the system goes into emergency heat mode?
17. What is a fair weather switch?
18. What is a TXV, and how does it work?
19. Why is the heat pump indoor thermostat different from other thermostats?
20. What are the two stages in the cooling mode?
21. What are the two stages in the heating mode?
22. When servicing a heat pump, whose directions should always be followed?

Chapter 15

Communication and Professionalism

The success of a service technician depends on the customer, and the customer's perception of the service technician. The service technician must always try to get along with customers, no matter how difficult they may seem. It is also necessary to continue to improve relationships with all customers, existing and new.

In order to be successful, the service technician must be a mechanic, a problem solver, and a salesperson. However, the most important aspects a service technician must have are a good attitude and a respect for other people.

CUSTOMER RELATIONS

Customers are people who deserve recognition, respect and assurance when they are faced with a problem. They need to know that their problem will be solved as quickly, effectively and fairly as possible. If the service technician treats the customer courteously and shows an interest in the customer's problem, chances are, the customer will respond positively.

It is important to remember that the customers paid a lot of money for the heating and air conditioning systems in their homes, and they expect the systems to operate properly. The heating and air conditioning systems were purchased for one basic reason: to provide comfort to the living space. The customer also expects the system to provide convenient and dependable operation. Consequently, when the system malfunctions, the customer expects the trouble to be corrected promptly and economically.

The customer must have confidence in the service technician's ability to correct the problem. At the time of the service call, the customer may be frightened by a potential safety hazard, or angry that an expensive unit is now malfunctioning. When this occurs, it is necessary for the service technician to take the position of the expert, who can correct these problems, and get the system back in operation in a minimum amount of time and at the lowest cost to the customer. In this manner, the customer normally feels reassured.

EDUCATING THE CUSTOMER

After completing a new installation, it is wise for the service technician to show the customer how to get maximum comfort from the new equipment. The service technician should also show the customer how to keep the system in top condition. This enhances customer relations and also extends the life of the system. The service technician should then explain to the customer how to call for service if anything goes wrong.

The service technician may provide the customer with the following information after a new installation:

- Set the automatic thermostat at the desired comfort level, and leave it alone all year.

- Make sure registers and returns are not blocked by furniture or drapes.

- Keep doors and windows closed.

- Use the kitchen exhaust fan when cooking.

- Vent the clothes dryer.

- Change filters regularly.

On initial installations, it is important for the service technician to explain the warranty provisions to the owner. Warranties vary with the manufacturer and type of component, but some common warranty periods are:

- Parts and Most Components: 1 year if defective in manufacture.

- Compressors: 1 year with an optional 4-year extended warranty at extra cost.

- Heat Exchangers: 5 to 10 years, depending upon various installation factors.

As stated previously, the service technician should show customers how to care for their newly-installed systems. This entails showing the homeowner, and any other residents, how to replace furnace filters before they become clogged and damage the system. The service technician should explain to the customer that performing this simple maintenance procedure will help the system operate more effectively. This suggestion should be stated in a positive manner, rather than suggesting the customer's home is dirty.

In a similar manner, the service technician should give the customer a brief explanation of the thermostat and how to obtain maximum comfort for the living space. The service technician may want to suggest that the customer keep the blower control in constant operation to avoid unsatisfactory comfort levels. Suggestions like this foster good customer relations and result in a more satisfied customer.

SERVICE CALLS

When responding to a service call, it is important that the service technician look competent. This means appearing neat and professional. If uniforms are available, the service technician should be wearing one, or else be in other appropriate attire. It is also necessary for service technicians to have identification stating who they are and which company they work for, as many customers ask to see this identification.

As the service call begins, it is necessary for the service technician to ask the customer a series of direct questions in order to find out just exactly what the trouble is. Adopting the practice of asking who, what, where, when and why helps the service technician pinpoint the problem. The service technician may ask the customer the following questions in order to find where the trouble lies:

- "Can you tell me what kind of furnace trouble you experienced?"

- "When did you first notice this problem?"

- "Did it happen only once or has it happened several times lately?"

- "Has this same problem occurred before?"

- "Does this problem seem to occur on certain days or certain times of the day?"

Initial answers to the questions may be vague, so it then becomes necessary to pursue a more detailed answer. For instance, when the service technician asks, "What seems to be the trouble?", the customer may simply reply that the temperature in the living area is not comfortable. The service technician should then continue asking questions until the customer says, for example, that the living space is cold in the morning, but the room heats up as the day continues. In this manner, the service technician accumulates the information needed to find and fix the problem.

As soon as the service technician obtains the necessary information, assurance should be given the customer that the general area of the problem is known and that it can be fixed. Assurance is especially important if the customer smells gas and fears a fire or explosion. In this instance, the service technician should find the problem immediately and assure the customer that there is no danger and the problem will be corrected. It is important to not, however, specifically pinpoint the problem at this time. This is important, because if the diagnosis is wrong, the customer may think the service technician is either incompetent or trying to overcharge for service.

Sometimes the service technician may be tempted to be discourteous. For instance, it may be particularly irritating to receive an emergency call at 2 a.m. when the problem is quite minor and could have been fixed by the homeowner. These minor problems may include an incorrectly-set thermostat or a blown fuse, however, the service technician must never indulge in rude behavior or insult the customer's intelligence. While this may give temporary satisfaction, it is bad business to insult a customer in any way. And, the customer is paying the bill.

Occasions arise where the customer, for one reason or another, may be agitated or mad and attempt to blame the service technician. Whether the customer is justified or not, it is important for the service technician to always hear the customer out completely. This action displays common courtesy, and the service technician may just learn more about the problem by listening carefully. The customer may also have a valid complaint, or else not have all the information necessary to understand the problem. If this is the case, the service technician should take this opportunity to explain the situation in more detail. It looks good for the service technician to come across as reasonable and concerned about the customer's opinion.

If the service technician is going to be late arriving at a customer's home, a call should be placed to inform the customer of this. If the customer must wait all afternoon for a service technician to show up, it can be quite an inconvenience, and the customer will probably be irate by the time the service technician arrives. Once the service technician arrives, no excuses should be offered. The service technician should simply apologize for the delay and assure the customer that this is not a usual circumstance.

In some cases, customers may be highly critical of the heating or cooling system they own. The service technician should then clarify the areas that the customer does not seem to understand. If the customer is still dissatisfied, the service technician should relay this information to the person who sold the customer the equipment and have this person contact the customer. The service technician should inform the customer of this procedure. It is very important that the service technician take steps to ensure a positive experience for the customer. An unsatisfied customer has only negative impressions of a particular company, and the customer will spread these impressions to friends, neighbors and co-workers.

For the most part, however, customers are usually appreciative and friendly.

CLOSING THE SERVICE CALL

In closing the call, the service technician must assure the customer that the problem is fixed and should not recur in the future. At this point, the service technician should carefully explain exactly what the problem was and what was done to correct it. Any parts that were needed should be itemized on a bill so the customer knows the exact cost of the materials. It is also good practice to leave the defective parts with the customer. However, if the parts are under warranty and the customer is not charged, the parts must be returned to the dealer. If the customer has any questions concerning the bill, the service technician should take the time to cover each item listed.

The service technician must make sure the work area is left clean and neat; that is, all parts picked up, boxes or wrappings collected, wire ends or tape removed, floor swept, and smudge marks or oil on the furnace wiped off. Any parts that have been replaced should be left in one place and the customer advised where they are.

Finally, give the service technician should give the customer the company name and address, plus telephone numbers for day and night service. Many service companies print this information on stickers which can be attached to the furnace for handy reference. You can also include space for the date of the last service call.

PROBLEM SERVICE CALLS

Sometimes it is not possible for the service technician to fix the problem. If this is the case, the service technician should carefully explain to the customer why the problem cannot be fixed and what the next course of action is.

If a special part is needed in order to fix the unit, the service technician should check the availability of the part and then explain the situation to the customer. If the part must be ordered, the service technician should give a reasonable estimate as to when it might arrive and when the replacement can be made. If the part is in stock, the service technician should give the customer the choice of returning and replacing the part the same day or finishing the job on another day.

A situation may arise where the electrical circuits are overloaded, requiring either rewiring to the junction box or shifting some other appliances from the main supply voltage. This is a job for a qualified electrician and one that the service technician is not equipped or licensed to handle. Inform the customer of the situation, and suggest that a call be placed to the original home electrical contractor, or another electrician.

EMPLOYMENT

Service technicians usually either work for a service firm or dealer, or are self-employed. In any case, service technicians must keep a set of standards in order to represent the company (or themselves, if self-employed) in the best possible fashion.

APPEARANCE

The attitude and appearance of a service technician should immediately give the impression of being competent and professional. This includes being neatly groomed and wearing a fresh uniform. Each job should be approached in a businesslike manner so that the customer is confident in service technician's abilities.

CARE OF EQUIPMENT

Since the employer has spent a considerable amount of money in supplying the service technician with a truck, tools and service instruments, the service technician should take good care of them. Many of the test instruments supplied by the employer are delicate, and must be well cared for if they are to remain operative and useful. These tools and instruments should be stored in an orderly manner. Also, all instructions provided with the instruments should be followed. Taking care of tools and instruments shows a basic respect for other people's property, which is a trait of all valuable employees.

HONESTY

Honesty is essential, both in what a service technician says to the customer and how the service technician reports back to the employer. For instance, it may be tempting for a service technician to exaggerate a malfunction to a customer in order to increase the sale. Or, it may be tempting for a service technician to "borrow", for personal use, some of the parts or supplies stored in the employer's truck. However, lying and stealing cost the employer money and business—and dishonesty is always detected eventually.

SAFETY

The service technician must follow proper safety precautions in using all of the tools and equipment provided. This is to ensure no harm is done to the service technician, the customer, or the customer's home. The employer is liable for any damage caused by the service technician on the job or traveling between jobs, so the service technician should be conscious of the safety factor at all times. This safety extends to the way in which the service technician drives the employer's truck. As the employer's business name is usually printed on the truck, it is in the best interest of the service technician to drive the truck responsibly. Driving irresponsibly creates a bad impression for observers and a bad impression of the employer.

ACCURACY

Accuracy includes doing the job right the first time and examining all of the factors that may have caused the original complaint. Sometimes a service technician locates and replaces a malfunctioning component to temporarily have the system back in operation, when the true cause of failure has not been determined. In such a situation, the system will certainly fail again. Call-backs for incomplete or inaccurate jobs are expensive to the employer, and they also damage the customer's opinion of the service technician.

FOLLOWING THE SCHEDULE

Most service technicians must perform a specific number of service calls in a day. Therefore, it is important for the service technician to be on time at a customer's house, and, barring any emergencies, perform all service calls on the schedule.

WRITTEN COMMUNICATIONS

Most service firms or dealers require written reports at the end of each day or each week, to keep track of the activities of a given service technician. As a result, it is easy for the supervisor to evaluate how a particular technician is performing on the job. Examples of reports a service technician might be required to file are listed below.

Inventory Reports. An inventory of spare parts and supplies must be carried on the truck. This list is usually checked daily or weekly at the main office. The service technician records information on the inventory report, such as which parts and supplies are used at which job. The parts and materials used can then be replaced, ensuring adequate supplies are always available to the service technician.

Loss or breakage of the tools and instruments assigned to the truck should also be reported, along with a clear explanation of how the damage occurred.

Cash and/or Check Accounting. In many cases, the service technician must present a bill to the customer immediately upon completion of the service call. To do this, however, the service technician must know the proper charges, including the chargeable labor rate and the cost to the customer for materials used. The bill must be filled out completely, totaled accurately and presented to the customer. The customer may then pay for the call. The amounts collected must be reported back to the main office and tied to a specific job so the customer can be given proper credit.

Call Reports. Many organizations require their service technicians to furnish additional information on each service call. This report or survey lists the materials used, the cause of the service complaint and what was done to resolve the situation. The inventory and cash accounting reports also become part of each customer's permanent record, so that the service firm always knows what service was performed for a particular call.

Customer Profile. The service technician should be aware of extras that the customer may need or want in the future, such as air conditioners, humidifiers or electronic air cleaners. This information provides valuable sales leads for future sales calls and direct mail campaigns. Also, the service technician may receive a bonus for helping to sell a particular system.

Return Goods Tag. Parts which must be replaced while still under warranty are returned to the dealer. The return goods tag is connected to the component once the dealer has filled out all appropriate sections. One section is made of cardboard and is connected to the component being returned. Other sections are used for the dealer's and manufacturer's records. The return goods tag shows the place of installation, date replaced, part type and serial number, reason for failure, and other information which helps the manufacturer to identify the malfunction. The tag should have a wire or other reliable method of attaching it to the defective part so that the two will not become separated in transit to the original manufacturer.

Some manufacturers have a labor allowance which pays for a portion of the labor required to change the part. This usually applies to larger components. In this case, labor time must also be recorded on the tag.

SERVICE CALL SALES

As stated previously, the service technician should be alert to all sales possibilities, particularly replacement sales. For example, if the customer has a particularly old furnace or one which requires major service, the service technician may point out that it would be better to replace the furnace, rather than repair it. If the furnace is old, the service technician could explain how efficient new furnaces are, and how the old furnace may keep breaking down.

If the heat exchanger is cracked or badly rusted, the service technician should point out the safety hazard to the customer. However, the service technician must always tell the truth. The service technician must not deceive the customer or exaggerate the description of the problem. A good approach is for the service technician to show the homeowner the damage and then explain the effect it will have.

Customers have greater confidence in a service technician suggesting a replacement than they do in a salesperson. The service technician's knowledge can benefit the customer by preventing future problems and inconvenience.

RADIO/PHONE USAGE

Modern service trucks are normally equipped with a two-way radio or portable phone for constant communication with the main office. Additional routing is made via the radio or phone, and emergency calls are relayed to service technicians in the field. The radio is a very convenient link to the main office, and the main office can then keep track of the service technician's location. The radio or portable phone should not be used for personal business or conversations that do not have a direct bearing on the job.

SUMMARY

It is not really necessary for the service technician to understand why people act as they do, but rather to respond to their actions in a courteous and positive way. If the service technician asks a number of questions at the beginning, and then goes about solving the problem quickly, a satisfied customer normally results. The service technician must be courteous at all times, and listen carefully to anything the customer says. It is also important for the service technician to remember that a smile goes a long way towards making a positive impression.

Finally, the service technician must always explain to the customer every service that has been performed. The customer then feels assured that the system will provide the expected comfort level. The service technician must gain confidence, calm any fears, and leave the customer's residence with the system working properly. This results in happy customers and a successful business.

Index

THE INSIDE STORY.

WILLIS CARRIER WAS THE FIRST INSIDE GUY.

Following a hunch he had about the relationship between humidity and coolness, Willis Carrier had a groundbreaking idea. In 1902, he put his idea to practical use. He invented air conditioning. Suddenly, being inside was comfortable year-round.

SINCE 1902, WE'VE BEEN MAKING IT BETTER INSIDE.

What began with Willis Carrier's brainstorm and hard work has been carried on by Carrier people all across the country. Residential air conditioning and heating has become a comfortable way of life, and Carrier dealers have become the "Inside Guys" people count on. Innovations such as energy saving heating and cooling equipment, humidifiers, air cleaners and electronic comfort controls have reinforced Carrier's name as #1 in comfort from coast to coast.

THE BRAND INSIDE AMERICA.

Carrier is the brand that homeowners know best. Since 1902, American consumers have relied on Carrier products for efficiency and comfort. And that confidence is why more Americans choose Carrier heating and cooling products.

Carrier
We're The Inside Guys.